NUTRIENTS VALORISATION VIA DUCKWEED-BASED WASTEWATER TREATMENT AND AQUACULTURE

Promoter: Prof. dr. H. Gijzen
Professor of Environmental Biotechnology
UNESCO-IHE Institute for Water Education,
The Netherlands

Prof. F. El-Gohary
Professor of Wastewater Treatment Technologies and Reuse
The National Research Centre, Egypt

Co-promoter: Prof. dr. J.A.J. Verreth
Professor of Fish Nutrition and Aquaculture
Wageningen University, The Netherlands

Dr. ir. P. van der Steen
Lecturer in Waste Water Treatment
UNESCO-IHE Institute for Water Education,
The Netherlands

Awarding Committee: Prof. dr. D. Mara
University of Leeds, United Kingdom

Prof. dr. P. Denny
UNESCO-IHE Institute for Water Education,
The Netherlands

Prof. dr. ir. A.J.M. Stams
Wageningen University, The Netherlands

Dr. ir. J.B. van Lier
Wageningen University, The Netherlands

Prof. dr. F. Ali Nasr
Professor of Wastewater Treatment Technologies and Reuse
The National Research Centre, Egypt

NUTRIENTS VALORISATION VIA DUCKWEED-BASED WASTEWATER TREATMENT AND AQUACULTURE

Saber Abdel-Aziz Abdel-Salam Mohamed El-Shafai

DISSERTATION

Submitted in fulfilment of the requirements of
the Academic Board of Wageningen University and
the Academic Board of the UNESCO-IHE Institute for Water Education
for the Degree of DOCTOR
to be defended in public
on Wednesday, 7 January 2004 at 13:30 h in Delft, The Netherlands

CRC Press
Taylor & Francis Group
Boca Raton London New York

CRC Press is an imprint of the
Taylor & Francis Group, an **informa** business
A BALKEMA BOOK

First published 2004 by A.A. Balkema Publishers

Published 2020 by CRC Press
P.O. Box 447, 2300 AK Leiden, The Netherlands
e-mail: Pub.NL@taylorandfrancis.com
www.crcpress.com – www.taylorandfrancis.com

ISBN 13: 978-90-5809-656-2 (pbk)
ISBN 13: 978-1-138-47504-5 (hbk)
ISBN 13: 978-90-5808-956-4 (Wageningen University)

Visit the Taylor & Francis Web site at
http://www.taylorandfrancis.com

and the CRC Press Web site at
http://www.crcpress.com

List of abbreviations

AD	Apparent digestibility
AWWT	Anaerobic wastewater treatment
AIA	Acid-insoluble ash
ADC	Apparent digestibility coefficient
BOD	Biological oxygen demand
COD	Chemical oxygen demand
CFU	Colony forming unit
DO	Dissolved oxygen
DMD	Dry matter digestibility
ED	Gross energy digestibility
EC	Electric conductivity
FCP	Freshwater-commercial feed-fed fishpond
FCR	Feed conversion ratio
FD	Fat digestibility
FDP	Freshwater-duckweed-fed fishpond
FWP	Freshwater-wheat bran-fed fishpond
GBH	Gravel bed hydroponics
HRT	Hydraulic retention time
LOEC	Lowest observable effect concentration
N	Nitrogen
P	Phosphorus
PD	Protein digestibility
PER	Protein efficiency ratio
PhD	Phosphorus digestibility
TAN	Total ammonia nitrogen
TCP	Treated sewage-commercial feed-fed fishpond
TDP	Treated sewage-duckweed-fed fishpond
TKN	Total kjeldahl nitrogen
TMS	Trimethylsilyl
TP	Total phosphorus
TSS	Total suspended solids
TWP	Treated sewage-wheat bran-fed fishpond
SSP	Settled sewage-fed fishpond
SGR	Specific growth rate
UASB	Up-flow anaerobic sludge blanket
UIA-N	Un-ionised ammonia nitrogen
WSP	Waste stabilization pond

Abstract

Saber Abdel-Aziz Abdel-Salam Mohamed El-Shafai (2004) Nutrients Valorisation via Duckweed-based Wastewater Treatment and Aquaculture. Ph.D. Thesis, Wageningen University and UNESCO-IHE institute for water education, The Netherlands.

The costs of mainstream wastewater treatment technologies, both in terms of construction as well as in terms of operation, keep many small to medium sized communities back from utilising reclaimed water for unrestricted reuse purposes. These treatment technologies are not only expensive, but also do not allow nutrients recovery. The main purpose of this research study was therefore the development of a sustainable treatment scheme to recycle nutrients and water. Use of an integrated UASB-duckweed ponds system for domestic wastewater treatment and recycling of nutrients and water in tilapia aquaculture was investigated. Monitoring of the integrated UASB-duckweed ponds showed that the overall efficiency of the system for organic matter removal (COD, BOD and TSS) was not significantly affected by temperature. Effluent quality was 49 mg COD/l, 14 mg BOD/l and 32 mg TSS/l in summer (24-34 °C), in comparison to 73 COD/, 25 mg BOD/l and 31 mg TSS/l in winter (13-20 °C). Nutrient removal was significantly reduced in winter, which shows that the nutrient removal processes are more temperature dependent than the BOD and COD removal mechanisms. Residual values of ammonia, TKN and total phosphorus were 0.4 mgN/l, 4.4 mgN/l and 1.1 mgP/l in summer, in comparison to 10.4 mgN/l, 12.2 mgN/l and 2.7 mgP/l in winter. The faecal coliform count in the final effluent was $4.0 \times 10^3 \pm 3.7 \times 10^3$ cfu/100ml and $4.7 \times 10^5 \pm 5.5 \times 10^5$ cfu/100ml in summer and winter, respectively. A nitrogen mass balance in summer showed that the most important removal mechanism for nitrogen in the duckweed ponds was plant uptake (80.5%), followed by denitrification (14.8%) and sedimentation (4.7%). Ammonia volatilisation was negligible in all ponds. The nitrogen uptake ranged from 4 to 4.9 KgN/ha/d in summer and from 1.2 to 1.5 KgN/ha/d in winter. For phosphorus it ranged from 0.86 to 0.97 Kg P/ha/d in summer and from 0.27 to 0.32 Kg P/ha/d in winter. The duckweed production rate was 126-139 kg dry matter/ha/d in summer and significantly decreased in winter to 31-36 kg dry matter/ha/d. These results showed that the UASB-duckweed ponds system is technically appropriate for sewage treatment in small communities and rural areas and provides marketable by-products (duckweed biomass).

Nutritional value of duckweed biomass harvested from the duckweed ponds was evaluated for raising tilapia using semi-batch experiments. The duckweed (*Lemna gibba*) was evaluated as sole feed source for Nile tilapia at temperature range of 16-25 °C using fresh and dry duckweed at different feeding rates. Statistical analysis showed significantly ($p < 0.01$) higher specific growth rate (SGR) for tilapia fed with fresh duckweed than for the parallel treatments fed with dry duckweed. In case of fresh duckweed, the SGR of tilapia fed on 10% feeding rate had a significantly ($p < 0.01$) lower value than was observed for 25% and 50% feeding rate. There was no significant ($p > 0.05$) difference between 25% and 50%. In case of dry duckweed no significant ($p > 0.05$) differences were detected between the 25% and 50% feeding rate. Results of the nitrogen mass balance showed that the nitrogen content of dry sediment (solid waste) ranged between 4% and 10.5%. In case of optimal feeding rate (25% fresh duckweed) the value was 4.5%. This amount of nitrogen represented 8% of the dietary nitrogen input while nitrogen recovery by fish represented 26-28% of the total nitrogen input.

Use of both treated sewage and duckweed biomass from the UASB-duckweed pond system was evaluated in rearing Nile tilapia (*Oreochromis niloticus*) in ponds. Comparison of fresh duckweed with local fish feed ingredient (wheat bran) and commercial fishmeal based diet was investigated. Three different water sources; freshwater, treated effluent of the UASB-duckweed ponds system and settled sewage were investigated. The results of growth performance proved that, in case of freshwater ponds, the SGR of tilapia fed on fresh duckweed was significantly ($p < 0.01$) higher than the SGR of tilapia fed on wheat bran, but lower than the SGR obtained with fishmeal-based diet. In case of treated effluent fed ponds, there was no significant difference ($p > 0.05$) between the three feeds. This indicates the potential of duckweed to replace wheat bran as supplemental fish feed. For all three feeds the treated sewage fed ponds provided significantly better SGR than in the freshwater fed ponds. A negative net yield (-0.16 ton/ha/year) was obtained in a settled sewage-fed fishpond, due to high mortality in fish biomass (60% in the adult fish and 38% in the fry). This mortality was attributed to an average un-ionised ammonia concentration of 0.45 mg N/l during the autumn (mortality period).

Effects of chronic un-ionised ammonia nitrogen (UIA-N) concentrations on the growth performance of tilapia (*Oreochromis niloticus*) fed on fresh duckweed (*Lemna gibba*) were investigated. Statistical analysis of SGR showed no significant differences between the SGR of tilapia in the control (0.004 mg UIA-Nl-1) and the SGR of tilapia exposed to 0.068 mg UIA-N l^{-1}. The SGR of tilapia exposed to un-ionised ammonia nitrogen over 0.068 mg UIA-N l^{-1} (0.144, 0.262 and 0.434 mg UIA-N l^{-1}) was significantly reduced. The maximum no observable effect concentration was 0.068 mg UIA-N l^{-1}, while the lowest observable effect concentration was 0.144 mg UIA-N l^{-1}. No fish mortality was detected in this trial, even at 0.43 mg UIA-N l^{-1}. Such a concentration caused serious mortality in the experiment with a settled sewage-fed fishpond (see above). The mortality in that pond may possibly be due to synergistic effects of other water quality parameters, like sewage bacteria count and low dissolved oxygen at dawn, to the toxicity of UIA-N. The results of microbial quality of tilapia reared in four faecal contaminated ponds support this. In spite of absence of any significant differences between the pond water microbial quality in treated sewage ponds and settled sewage fed pond, significantly higher contamination was observed in fish organs in the latter. This strongly supports the hypothesis of interrelated synergistic effects between microbial quality and some physicochemical parameters like ammonia, nitrite and/or low dissolved oxygen at dawn. The poor water quality in settled sewage-fed pond resulted in faecal coliform counts in fish organs of about one log_{10} higher than in the treated sewage fed ponds. Therefore pre-treatment of sewage to remove pathogens, organic matter and nitrogenous compounds is therefore recommended before use in aquaculture.

Apparent digestibility coefficients of duckweed (*Lemna minor*) were measured for Nile tilapia (*Oreochromis niloticus*). Tilapia was fed on four iso-nitrogenous treatment diets in addition to control diet. Effect of partial replacement of control diet components with 20% and 40% dry duckweed and 20% and 40% fresh duckweed was investigated. Specific growth rates of tilapia were 1.51±0.07, 1.38±0.03, 1.31±0.06, 1.44±0.02 and 1.33±0.05, in control and treatments 1 to 4. All of the treatment diets provided good values for feed conversion ratio (FCR, 0.98-1.1) and protein efficiency ratio (PER,

2.49-2.78). All the treatment diets had high protein digestibility (78%-92%) and high-energy digestibility (78.1%-90.7%). It is therefore possible to use duckweed as replacement of fishmeal and other plant ingredients in fishmeal-based diets at 20% and 40% inclusion. For 20% duckweed inclusion there was no significant difference between dry and fresh duckweed for the apparent digestibility coefficients of dry matter, protein, fat, phosphorus and energy. However, in case of 40% inclusion the fresh duckweed had significantly lower digestibility coefficient values for dry matter, protein, fat and phosphorus. The milling process might have had positive effects on the apparent digestibility of dry duckweed through destruction of cell walls and decreasing the particle size, consequently increasing the specific surface area for enzyme reactions. Body composition showed that tilapia fed on diets with duckweed has significantly higher phosphorus and protein content and a significantly lower lipid content. This might be attractive for the consumers. In conclusion treated domestic sewage and fresh duckweed (*Lemna gibba*), from a UASB-duckweed ponds could be used in semi-intensive tilapia pond culture. The duckweed could be used in intensive tilapia culture either as partial substitute of fishmeal or complete substitute of some plant ingredients.

Table of Contents

Chapter One
Introduction and Literature Review

Introduction and Literature Review

1. Introduction

The world is facing a variety of interrelated problems such as water scarcity, lack of sanitation, population growth, unlimited food production capacity, energy availability, unemployment and urbanisation. There is a need for rapid economic development to increase food production for feeding the fast expanding human population and supply them with basic requirements. A combination of population increase and increasing demands on natural resources as a result of agricultural and industrial development in recent years has focused governmental attention on the need to retain life support systems and the amenities of land, energy and water for future generations. Many ecological disasters that have occurred as a result of non-sustainable use, abuse and misuse of natural resources, have clearly demonstrated that long-term sustainable development can be achieved only through sound environmental management. The need to learn from the experience of ecological-unsustainable development and to prevent its repetition on a global scale has been generally accepted.

The major problems are lack of safe drinking water, safe sanitation and food security. One-third of world population lives in countries suffering from moderate to high water stress, where water consumption is more than 10% of renewable freshwater resources. Some 80 countries, constituting 40% of the world population, were suffering from severe water shortages by the mid 1990's and it is expected that in less than 25 years two-thirds of the world's people will be living in water stressed countries. In Africa, at least 13 countries suffered water stress or water scarcity (less than 1700 m^3/capita/year and less than 1000 m^3/capita/year) in 1990 and the number is expected to double by 2025 (UNEP, 2002). The growing expansion in water supply without equivalent expansion in sanitation services especially in rural areas leads to contamination of surface and ground water causing widespread water quality deterioration. The total African population with access to improved sanitation was 60% in 2000, with an average 84% in urban areas and 45% in rural areas (WHO and UNICEF, 2000). Sanitation coverage in rural areas is less than in urban settings where 2 billion people of those lacking adequate sanitation live in rural areas. A rapid action is now urgently needed to establish new credible targets for water resources and sanitation by refocusing the efforts of both governments and development communities.

It is now widely appreciated that the so-called protein gap is a fact and that persistent and widespread malnutrition in many developing countries is due to a lack of both sufficient food and sufficient good quality protein. Animal protein contains all the essential and non-essential amino acids (20 amino acids) needed for healthy growth in human beings. Lack of animal protein and its non-affordable prices for the majority of people make it urgent to increase its production, decrease its production costs and focus on other available animal protein sources. Fish emerges as one of the major protein sources for many developing countries. Fish is particularly valuable nutritionally, because in addition to having protein with balanced amino acids, it contains vitamin B12 and it is low in cholesterol and saturated fat.

Aquaculture is becoming an increasingly important source of fish products as harvests from natural waters decline world-wide due to over fishing and water pollution, and due to increasing demand for animal protein. There are currently many attempts to increase fish aquaculture production at all levels, notably by the efforts of the Food and Agricultural Organisation (FAO) and other international organisations. Currently, most fish aquaculture farms are established in outdoor pond systems where water is diverted from a nearby lake or river for a single use in the pond systems and then released again to the environment. Though use of water in aquaculture is often considered as "non-consumptive" there are significant losses due to seepage and evaporation, depending on the soil properties and climatic conditions. Furthermore competing uses of surface water, such as drinking water supply and irrigation might cause restrictions for aquaculture especially in water scarce regions. Water that is unsuitable for drinking and irrigation, such as wastewater and saline water represent a potential substitution of fresh water in aquaculture. In this way aquaculture farms could simultaneously contribute to effective environmental management and to the socio-economic well being of human populations. It means that when domestic wastewater and farm wastes are used for fertilising or feeding the fishponds, this not only constitutes an inexpensive mean of waste disposal, but it also contributes to effective recycling of resources to produce food and fodder. Further more fishponds often serve as a source for crop irrigation, watering livestock and for household uses, which means a higher regional water efficiency.

Domestic wastewater contains three main elements that could be considered for recovery and reuse schemes. Organic carbons could be converted into methane gas via anaerobic treatment, and be reused as a source of energy within the treatment facility. Nitrogen and phosphorous, which represent the major nutrients in wastewater could be transformed into high quality protein biomass, via either aquaculture or agriculture or both. Reuse of wastewater in aquaculture and agriculture could represent a better and preferred method of wastewater disposal and should form an integral part of water and nutrient resources planning. Recent efforts are being directed to stimulate the development of using wastewater as feed supplement in fish farming. The direct use of wastewater in fish aquaculture is applied in some Asian countries. But this practice poses a number of limitations. Major limitations of direct use of wastewater in fish aquaculture are caused by the presence of high number of human pathogens, social aspects, market value of fish and toxic parameters, mainly NH_3 and nitrite. Alternatively many of the potential benefits to be gained by treating and recycling organic waste and nutrient resources in wastewater using aquatic plants-based systems have been well demonstrated and documented and should lead to an increase in such uses in the future.

Sustainable wastewater treatment comprising an Up-flow Anaerobic Sludge Blanket (UASB) reactor as primary treatment unit followed by duckweed ponds is considered a cost-effective way of energy and nutrients recovery and water recycling. In case of anaerobic treatment there is no need to supply mechanical aeration that is very expensive energetically and equipment wise. Anaerobes convert biodegradable organic matter into biogas (methane gas), which could be collected and reused as energy source in case of anaerobic reactors, something, which is more complicated in case of anaerobic ponds. Anaerobic degradation of the complex nitrogen and phosphorus containing organic compounds into smaller soluble inorganic ones available for the duckweed may enhance a good biomass yield (Alaerts *et al.*, 1996). Likewise, decreasing the organic load via anaerobic pre-treatment may enhance removal of pathogenic bacteria in the subsequent stabilisation ponds (Dixo *et al.*, 1995) like algae-based ponds and duckweed-based ponds. Duckweed is one of the potential aquatic macrophyte plants for wastewater treatment. The duckweed has a high nutrient absorbing capacity, a high growth rate and is extremely tolerant to the environmental conditions. Nutritionally, duckweed has a high nutritive value represented by high protein content and low fibre content. Water leaving the system could be used for fish aquaculture and/or crop irrigation.

Presently, application of duckweed in fish aquaculture attracts attention of farmers in different countries of East Asia. Fish aquaculture using sewage-duckweed lagoons has a weak scientific base, and much of the research so far has been directed to the system to achieve high fish production, maximum nutrient recovery and good effluent quality. A major challenge for future sewage-duckweed-fish farming integrated systems is to improve our understanding of critical ecosystem food chain control and process mechanisms. This could provide possibilities to optimise system performance in terms of fish production, water quality and nutrient recycling. Investigating the UASB-duckweed ponds system for sewage treatment and reuse of both treated effluent and duckweed biomass in tilapia pond aquaculture was the main focus of this study. So far the effect of duckweed form (fresh or dry) and feeding rate on water quality and fish yield has not been studied extensively and therefore was subject in this research. Evaluation of duckweed in tilapia pond aquaculture was proposed by comparing it with wheat bran and fishmeal-based diet. Ammonia concentration and faecal coliform count in the pond were also studied since these might suppress the growth yield and microbial quality of tilapia. Partial or complete replacement of fishmeal and scarce plant protein sources with duckweed could be of considerable economic advantage even if this would be accompanied by a moderate reduction in fish feed quality and feed utilisation efficiency. Since feed formulation should be based on nutrient availability, reliable data on the digestibility of different ingredients are important and that is why measurement of apparent digestibility coefficients for feed ingredients is important. Such data are available for some commonly used ingredients. The apparent digestibility coefficients of duckweed have not been measured yet and need to be examined.

2. Literature review and background

For a large part of the world population, one of the greatest environmental threats to health remains the continued use of untreated water and lack of sanitation. At the beginning of the UN drinking water supply and sanitation decade (1980-1990) an ambitious target of universal access to safe drinking water and affordable sanitation for all was set for the end of the decade. This target was subsequently extended to the year 2000. Efforts of WHO and UNICEF and other international organizations achieved clean water supply for more than 800 million and safe sanitation for around 750 million between 1990 and 2000 (WHO and UNICEF, 2000). Despite the intensive efforts of many national and international organizations, 1.1 billion people (one-sixth of the world's population) still remain without access to improved sources of water and 2.4 billion (two-fifth of the world's population) are living in unhealthy environments due to lack of access to appropriate sanitation services. 0.03 billion out of 1.1 billion people without access to safe drinking water, live in Africa (WHO and UNICEF, 2000). Lack of access to safe drinking water and safe sanitation resulted in hundreds of millions of cases of water-related diseases, and more than 5 million deaths every year with 3 million only in Africa (UNEP, 2002). About 4 billion cases of diarrhoea each year causes 2.2 million deaths every day in developing countries most of them children, die every day from disease associated with lack of safe drinking water and inadequate sanitation (WHO and UNICEF, 2000).

Will there be enough water to grow food for the almost 8 billion inhabitants expected to populate the earth 2025? (Rosergant, *et al.*, 2002). The water crisis will have severe consequences on global food production. By 2025, water use is expected to increase by 40%; 17% more water will be required for food production to meet the needs of the growing population and 23% increase in other uses. The major cause of growing water demand is rapid population growth, which increases food demand in many developing countries. There is no doubt that efficient use of existing water resources and exploration of new resources are needed to achieve this goal. Over the last decade, there have been a series of international summits held by the United Nations and its specialized agencies, at which they have discussed the need for achievement of greater progress in poverty reduction and sustainable development. Sustainable water resources management could keep food demand at affordable prices to poor people. Water resource management and exploration of new resources could be established through implementing decentralised sanitation and reuse concept. Decentralised concept will decrease the plenty of water that is being used to transport wastes in the large pipelines centralised facility. The reuse concept will represent new source of water and nutrients for crop production.

2.1. Environmental Situation in Egypt

In most parts of Africa the chronic shortage of water has sever consequences for the people of the continent. Despite the commendable activities of many governmental and charitable organisations millions of people will continue to suffer as the deficient in quantity and quality of water supplies dwindle. This imposes the decision makers and governmental authorities in Egypt to develop a national strategy for optimal utilisation of all available water resources including reuse of wastewater. Before examining wastewater treatment approach and its role in fish aquaculture and agriculture practices it is appropriate to review recent trends in the current environmental situation in Egypt. The non-conventional water resources in Egypt such as treated sewage could represent a key element in the process of national development in the near future.

The population of Egypt was 36 million in 1960, 56 million in 1990 and reached 67 millions by the year 2003. Due to shortage in wheat and oil seed production, Egypt imports two-thirds of the national wheat and vegetable oil demand. The nutrition gap between domestically produced food and national consumption in Egypt, which must be filled by food imports, was estimated at 40% in 1994 (Abdel Mageed, 1994). Nowadays (2003) this gap represents 15 million tons of agricultural products per year. The target of the land reclamation plans of the ministry of public water works and water resources and the ministry of land reclamation was set at 1.36 million ha of new land to be cultivated by the year 2000. In the plans the irrigation water for 0.2 million ha is foreseen by exploiting ground water and of 0.6 million ha by reducing conveyance losses of the Nile catchments as it passes through the Sudd, southern Sudan. The remaining 0.6 million ha depend on rationalisation of water use in the 3.9 million ha presently irrigated. This could be achieved through more efficient on field use and re-utilisation of drainage water. Also the reuse of wastewater after treatment could make a significant contribution in the availability of irrigation water. Initial steps have been taken to establish pilot projects on the use of treated wastewater in agriculture and to address several environmental issues associated with improved water management.

The total municipal sewage in Egypt is about 3.5 billion cubic meter per annum, which is three times the amount of annual produced rainfall (Table 1). In addition the nitrogen content of this amount of sewage is estimated at 0.14 million ton. This amount of nitrogen could fertilise 0.9 million ha of cotton or wheat. The price of an equal amount of the inorganic fertiliser is $ 30 million. The net water requirements of cotton and winter wheat are 12,240 m^3/ha/season and 10,800 m^3/ha/season, respectively. By using drib irrigation technique the gross water requirements of cotton and wheat will be 14,076 m^3/ha/season and 12,420 m^3/ha/season. Based on the previous data, reuse of domestic sewage in Egypt could add 0.25 million ha of cotton or 0.28 million ha of wheat, which represent 6%-7% of the present 3.9 million ha cultivated area. Depending on the location of the areas to be reclaimed and the location where wastewater is available in suitable quantities and quality, wastewater effluent may be directly used with or without mixing with fresh water. The Salam canal project for the reclamation of 0.4 million ha is located in the Eastern Nile Delta and in the Sinai (0.2 million ha in Al-Arish area), and the Umum reuse project for the reclamation of 0.2 million ha is in the Western Nile Delta. In both cases, blending of the wastewater with fresh Nile water is required and also planned by the ministry of public works and water resources. Although reuse of wastewater can be considered as a fast and economic solution to compensate water shortage in reclamation projects, it still needs investments in infrastructure.

Egypt's share in the River Nile water is about 55.5 billion cubic meters per annum, Table 1. This amount covers all the economic sectors in Egypt in addition to potable water supply. The prolonged drought period of the last eight years in Africa has affected the discharge of the River Nile upstream of the Aswan High Dam. During this period, live storage of the dam has been sufficient to supplement the Nile discharge to the normal water level required for agriculture, industry and domestic use. However, if the drought period persists for a number of years, the live storage may be exhausted and an acute water crisis may develop. Under these circumstances a water shortage of 20%, could be acute and only water resources management can be implemented to minimise the damage of such a shortage in terms of agricultural production. The long-term effects of diverting fresh Nile water to land reclamation projects on the account of the water budget of the old valley can be minimised by using wastewater in the reclamation of these lands. The WHO scientific group on health aspects of reuse of treated wastewater for crop irrigation and aquaculture emphasised the need to ensure, that appropriate controls are incorporated in schemes for wastewater reuse to protect both the environment and public health. Furthermore, protective measures for water and crops are to be made available (WHO, 1989).

Table 1: Water resources in Egypt (in billion cubic meters per annum).

| References | Rainfall | River Nile | Ground water | | | Municipal sewage | | |
			Storage	Available	$R^{1)}$	Raw	Treated	Reused
Abdel-Mageed, 1994	15[*]	55.5	$6000^{2)}$	2.6	4.5	3.9	0.2	Not available
El-Gohary (2002)	1[**]	55.5	Not available	4.8	7.5	3.5	1.6	0.7

[1)] Ground water recharge, [2)] Usable and renewable ground water, [*] Total rainfall, [**] Harvestable rainfall

The Egyptian government promotes the development of comprehensive health services, the prevention and control of diseases and the improvement of environmental conditions. The technologies available to treat wastewater to a standard high quality for safe discharge into waterways or reuse have generally led to the construction of relatively sophisticated wastewater treatment plants. Activated sludge and oxidation ditches represent 58% of the technologies implemented and 72% of the total wastewater treatment capacity (El-Gohary, 2002). Currently the total volume of wastewater is 3.5 Billion Cubic Meter (BCM); 1.6 BCM receiving treatment and 0.7 BCM from the treated effluent is being used (Table 1). In Greater Cairo, the construction and rehabilitation of five wastewater treatment plants (WWTP) have been completed. The capacity of the Gabal El-Asfar secondary treatment plant was 10^6 m^3/day at the first stage. This capacity increased to 3×10^6 m^3 /day by 1995 and serves 12 million people. A secondary WWTP with 0.33×10^6 m^3 /day treatment capacity, exists at El-Zenein and a 0.4×10^6 m^3 /day primary treatment plant exists at Abu-Rawash. Two other treatment plants exist at Berka (0.6×10^6 m^3 /day to primary standard) and Shoubra El-Kheima (about 0.6×10^6 m^3/day). However, in many cases these have proved beyond the capabilities of local authorities to maintain or operate. As a result, expensive high technology plants have fallen into disrepair and in many instances, have produced effluents, which do not comply with the general standards recommended by Egyptian legislation.

In Egypt, Cairo has recently focused on a new major sewage system in which consideration is given to the construction of sewage treatment plants that would yield 2 to 3 billions cubic metre of water annually of a quality that is adequate for crop production. El-Zenein WWTP as mentioned is one of five rehabilitated wastewater treatment plants in Greater Cairo. It has been constructed through the greater Cairo wastewater treatment project in 1990 (Khalil and Hussein, 1997). The plant employs activated sludge technology and treats about 330,000 cubic metres per day. The primary treatment of wastewater in this plant includes screening, grit removal, pre-aeration and then primary clarification by separating about 75% of settleable solids from the wastewater (primary sludge). In the secondary (final) treatment, more than 95% of the solids in the raw wastewater are removed using activated sludge. The accumulated sludge represents a big challenge to the plant authority. This plant provides effluent with a quality that does not comply with the discharge standard, since ammonia nitrogen and phosphorus concentrations reach about 28.3 and 16.8 mg/L, respectively.

Table 2: Greater Cairo wastewater treatment plants (Kell *et al.*, 1993; Khalil and Hussein, 1997)

WWTP	Gabal El-Asfar	El-Zenein	Abu-Rawach	Berka	Shoubra El-Kheima
Treatment stage	Secondary	Secondary	Primary	Primary	Secondary
Working capacity (m^3/d)	3×10^6	0.33×10^6	0.4×10^6	0.6×10^6	0.6×10^6

Lagoon treatment is a simple, low cost method to handle treatment of domestic wastewater and is most often used by small communities. Daqahla wastewater treatment plant, implemented in Egypt in 1989, includes two anaerobic ponds and one aerated lagoon followed by three maturation ponds (El-Sharkawi *et al.*, 1995). In spite of serving 80% of the design population (7500), Daqahla waste stabilisation pond is loaded with an actual flow (833 cubic metres per day) much higher than the design flow (520 cubic metres per day). This is caused by high average daily flow rates in addition to high strength wastewater. This produces effluent quality does not comply with the standard.

In the city of Suez, about 400 cubic metres per day of raw sewage from the city is diverted and treated by two parallel pond systems (Shereif *et al.*, 1995). The sewage experimental station occupies about 14 ha to the west of the city of Suez, adjacent to the city sewage treatment plant. Oxidation ponds and aquaculture facilities have been constructed on 5.3 ha and the remaining area is being used for irrigation application. The first treatment system consists of conventional waste stabilisation ponds, which include, anaerobic, facultative and maturation ponds with a total residence time of 21 days. The second system includes a series of four plankton ponds with a total residence time of 26 days. The results obtained indicated that the treatment plants provide effluent that conforms to the WHO standards. The effluent is successfully used to grow two types of local fish and subsequent reuse of fishponds effluent in crop irrigation is without health risks (Easa *et al.*, 1995; Shereif and Mancy, 1995; Shereif *et al.*, 1995).

Gravel Bed Hydroponics (GBH), a constructed wetland system, is being used in Egypt for secondary treatment of wastewater (Williams *et al.*, 1995; Stott *et al.*, 1997). A field scale GBH constructed wetland system has been in operation at Abu Attua, Ismailia, Since 1988. The GBH system is designed as a horizontal sub-surface flow system with inclined gravel beds planted with local reeds (*Phragmites australis*) and received treated wastewater from conventional trickling filter. The effluent of GBH with faecal coliform count of $4.9 \times 10^3 / 100$ ml does not satisfy WHO guidelines for unrestricted irrigation.

2.2. Sustainable Wastewater Treatment

Sustainable development is a development that simultaneously creates or preserves the vitality of society, economy and the environment. The triple bottom line aspects of sustainability cover environment, economy and society. Environmental objectives of sustainability are the protection and conservation of resources on local, regional and global scale. The economic goals include cost reduction and recovery while social objectives are improvements in the quality of life and food security for the society. Sustainable wastewater treatment should meet these requirements via reducing the cost of the treatment, recovery of the cost by recycling of the nutrients and water for the benefits of society and the environment. Recycling of the nutrients from wastewater treatment back to aquaculture/agriculture practices is essential for sustainability, as these nutrients are needed as fertilisers. Artificial fertilisers are energy-intensive products causing depletion of resources and pollution. In a conventional activated sludge wastewater treatment system nutrients are accumulated in sludge or flow into the water streams. Consequently, these nutrients are no longer available for aquaculture/agriculture uses but instead may cause eutrophication. Reuse of these nutrients is essential to ensure fertility of soil in the future.

2.2.1. Anaerobic pre-treatment

Contrary to centralize, mechanized and energy intensive activated sludge wastewater treatment systems, anaerobic treatment presents an energy efficient treatment that can be applied at any scale. Also the anaerobic treatment process is increasingly recognised as an important step of an advanced treatment technology for environmental protection and resource preservation and it represents, combined with other proper post-treatment methods, a sustainable and an appropriate wastewater treatment system for developing countries. Anaerobic treatment of sewage is increasingly attracting the attention of sanitary engineers and decision-makers. It is being used successfully for wastewater treatment in tropical countries, and there are some encouraging results from subtropical and temperate regions.

Anaerobic Wastewater Treatment (AWWT) has several advantages over conventional aerobic sewage treatment (Mergaert *et al.*, 1992). These advantages can be listed in the following points:

1-It has less sludge production. Also sludge has higher settleability and better dewatering characteristics.
2-It has less energy requirements (no aeration).
3- It has valuable methane gas production, which may substitute on site use of non-renewable energy.

4-It is capable of removing some compounds (e.g. chlorinated organics), which the aerobic process does not remove.

5-It has lower investment, operation and maintenance costs.

The anaerobic treatment is a mineralisation process, which converts organic carbon into carbon dioxide and methane, nitrogenous compounds into ammonium salts and organic phosphates into inorganic phosphates. Seghezzo *et al.* (1998) reported that the Up-flow Anaerobic Sludge Blanket (UASB) reactor appears today as a robust technology and is by far the most widely used high-rate anaerobic process for sewage treatment.

A review of published works of UASB reactor performance gives us a good idea about the advantages of this reactor. The results obtained by Barbosa and Sant Anna (1989), showed the possibility of starting up the UASB reactor for raw domestic sewage treatment without seed inoculation. The application of this technology to domestic sewage is rapidly growing in developing countries (Draajer *et al.*, 1992; Schellinkhout and Collazos, 1992; Vieira and Garcia, 1992; Vieira *et al.*, 1994; Seghezzo *et al.*, 1998).

Dixo *et al.* (1995) found that the UASB reactor, with a HRT of 7 hours, is an efficient primary treatment alternative to an anaerobic pond in Waste Stabilisation Pond (WSP) series receiving strong domestic wastewater. The UASB has an important role in removing total suspended solids, biological oxygen demand and intestinal nematode eggs but is deficient in removing pathogenic bacteria and nutrients. The advantages of UASB over an anaerobic pond can be described as follows.

- The advantage is complete mixing flow in the UASB reactor compared to horizontal incomplete mixing in the pond.
- The UASB has good substrate-biomass contact.
- Removal of nematode eggs in the UASB reactor as a result of filtration and aggregation in the sludge granules verses inefficient sedimentation mechanism in the anaerobic pond.
- The UASB has good gas-liquid-solid separation.
- The biogas is trapped in the UASB reactor and could be subsequently reused or disposed of in a controlled manner.
- In ponds methane gas is released in the atmosphere and contributes to greenhouse gas emission.
- Sludge removal from the UASB reactor is a simple process while in the pond it is carried out as a batch removal process or using raft mounted pumps.
- The volume, detention time and area requirements of the anaerobic pond are 6.5 to 10 times more than in the UASB reactor.

2.2.2. Post-treatment of anaerobic effluent

Nutrient rich effluent from the UASB reactor should be post-treated not to remove but to recover and recycle these nutrients. The UASB reactor followed by oxidation ponds reduces the land requirements of stabilisation ponds and decreases sediment accumulation. High rate algae ponds represent a type of oxidation ponds adapted to high growth rate of algae. The high rate algae pond represents one of the most popular treatment methods and produces large amounts of algae biomass. Approximately 120 metric tons per year of dried algae-bacterial biomass containing between 40%-50% protein was expected to be harvested from a hectare of two-stage pond system while treating 600-1000 m^3 wastewater per day (Shelef *et al.*, 1982). It is noted that the high rate algae pond is 30 to 50 fold more productive than agricultural land in terms of yields of organic matter or protein. If this biomass could be harvested, it could safely substitute conventional plant protein animal feed ingredients such as soy meal in feeding of poultry and fish. In comparison with some aquatic macrophytes, algae have high productivity and protein content. The possibility of recovering the removed nutrients in the form of algae biomass however, represents a major limitation in wastewater treatment. The efficient recovery of the algal biomass is complicated and costly. Application of flocculation-flotation using chemicals (alum and ferric chloride), centrifugation and combination of these methods are used in algae removal from the effluent (Shelef *et al.*, 1982; Rodrigues and Oliveira, 1987; Sandbank and Shelef, 1987; Sandbank and van Vuuren, 1987).

The anticipated economics of high rate algae ponds with recovery of algal biomass are not very promising since the expenses associated with the micro algae harvesting and processing (mainly water extraction and the breakdown of the alga cell wall for animal consumption) make its profit marginal. Further research is required to develop alternative aquaculture methods for wastewater treatment and nutrient recovery. Current studies deal with a new wastewater purification system that may make the agriculture use of wastewater more convenient and the entire process more economic. Recycling systems based on the treatment of municipal wastewater with protein production using duckweed may represent a more cost effective and comprehensive solution (Cully and Epps, 1973; Oron *et al.*, 1984; Oron *et al.*, 1986; Oron *et al.*, 1987; Oron *et al.*, 1988; Hammouda *et al.*, 1995).

2.2.2.1. Duckweed-based pond

Crucial limitations of anaerobic treatment of domestic sewage relate to the absence of nutrient and pathogen removal and a combination of anaerobic pre-treatment followed by photosynthetic post-treatment via macrophyte-covered stabilisation ponds was proposed for effective recovery of nutrients and pathogen removal (Gijzen, 2001; Gijzen, 2002). Aquatic macrophytes play an important role in promoting both, nutrient transformations and recovery in aquatic macrophyte-based wastewater treatment ponds. The macrophytes promote nutrient transformation by acting as a biological active substratum for epiphytic microbial heterotrophs and nitrifiers. Through photosynthesis, radiant energy is stored as chemical energy in the form of biomass using inorganic carbon and nutrients. The utilisation rate of solar energy for wild plants is only 0.5%, and that of crops 0.5-1%. Algae and macrohydrophytes, however have a much higher utilisation efficiency in the range of 3-5% (Wang, 1991). This has attracted more and more farmers in developing countries to construct ponds to grow various species of macrophytes with high production rate. In addition to the role that aquatic macrophytes play to store radiant energy, it also acts as oxygen generator and carbon dioxide consuming machinery. Photosynthetic stored energy provides directly or indirectly all of the food energy consumed by animal and human population. Because of the high utilisation rate of aquatic macrophytes, these can play a vital role in biomass production from organic wastes produced by human population and domestic animals.

The direct conversion of ammonia into plant protein in duckweed ponds is a relatively energy efficient process compared to the energy-consuming nitrification process performed by activated sludge (Harvey and Fox, 1973; Oron *et al.*, 1987; Mbagwu and Adeniji, 1988; Zirschky and Reed, 1988; Oron *et al.*, 1988). Duckweed has high productivity, large nutrient uptake, easy handling and harvesting and extended growing period. In comparison to other aquatic plants duckweed has a high protein content and low fibre content (Table 3). The presence of a duckweed cover has also been reported to reduce mosquito breeding in the ponds. In contrast to duckweed, water hyacinths have low temperature sensitivity, high cost of plant harvesting and processing, high sludge accumulation in the pond, high evapotranspiration rate, extensive mosquito breeding and low protein content (Abdalla *et al.*, 1987; Kawai *et al.*, 1987; Santos *et al.*, 1987). Under optimal operating conditions duckweed ponds provide both, good quality secondary effluent that meets the WHO irrigation and aquaculture reuse criteria and annual yield of duckweed dry matter biomass of 55 ton/ha, dry matter (Oron, 1990).

Table 3: Nutritional value of different aquatic plants

Item	Crude protein[*]	Crude fibre[*]	Crude fat[*]	References
Algae	40-50	-	-	Shelef *et al.,* 1982; Rodrigues and Oliveira, 1987; Wong *et al.*, 1995
Water hyacinths	16	16.3	2.7	Abdalla *et al.*, 1987; Kawai *et al.*, 1987.
Water fern	22	3.7	1.5	Basudha and Vishwanath, 1997.
Duckweed	24-49	6.4-11.8	3.4-5.5	Culley and Epps, 1973; Rusoff *et al.*, 1980; Oron *et al.*, 1986; Mbagwu and Adeniji, 1988; Hammouda *et al.*, 1995.

[*] g per 100 g of dry matter

The concept of aquaculture implementation for wastewater treatment and recycling has received increased momentum over the past few years. It is believed that aquaculture methods can simultaneously solve environmental and sanitary problems and also present an economical way for food production. Smith and Moelyowati (2001) stated that the duckweed wastewater treatment systems are feasible for developing countries in hot climates to provide low-cost treatment of domestic sewage particularly in rural areas. In case of completely covered duckweed ponds, the water immediately below the duckweed layer contains high levels of suspended solids, and there is no diurnal fluctuation of oxygen within the bottom of the mat indicating that duckweed does not release much oxygen into the water column (Morris and Barker, 1977; Pokorny and Rejmankova, 1983). In contrast, Alaerts *et al.* (1996) estimated that aeration through the duckweed mat provides 3-4 g O_2/m^2/day, which is higher than oxygen transfer through an uncovered surface. These amounts of oxygen are likely to have a role in the treatment processes and will be consumed in BOD removal by heterotrophic bacteria in ponds.

Nutrients removal

The duckweed is a nutrient recovery machine that converts these nutrients into plant protein. In sewage fed fish farming ammonia becomes an important parameter, because of its toxicity to fish. Published data show that duckweed and other aquatic plants exhibit a preferential uptake of ammonia over nitrite and nitrate since the ammonia is transformed directly into plant protein rather than having to be reduced first as in the case of nitrate (Ferguson and Bollard, 1969; Nelson *et al.*, 1981; Porath and Pollock, 1982). Körner and Vermaat (1998) reported that 75% of the nitrogen and phosphorus removal was via plant uptake. The other 25% was due to non-duckweed related processes like, nitrification/denitrification through the duckweed attached biofilm and sedimentation. In a laboratory-scale experiment Al-Nozaily *et al* (2000) concluded that the main nitrogen removal is through plant uptake, providing 4.55 Kg N/ha/d. In a full-scale experiment, Alaerts *et al.* (1996) demonstrated that the duckweed sewage stabilisation pond system achieved 74% and 77% removal of nitrogen and phosphorus, respectively (Alaerts *et al.*, 1996). They concluded that the nitrogen balance in the system consists of 44% nitrogen uptake by duckweed, 28% seepage, 10% sedimentation, 9% release in the effluent and 9% unaccounted for. On the other hand Vermaat and his colleague (Vermaat and Hanif, 1998) investigated that the nitrification/denitrification is held responsible for 50% of the nitrogen removal in a duckweed-based wastewater treatment system and 34% was recovered in plant biomass. Similarly Zimmo *et al.* (2000) reported 40% nitrogen removal by denitrification and ammonia volatilisation in duckweed batch experiment. In an integrated duckweed-algae-based system the nitrogen removal was attributed to 18% plant uptake, 8% sedimentation and 73% volatilisation (Van der Steen *et al.*, 1998).

Pathogen removal

The public health aspect is one of the main constraints for reuse of sewage in fish aquaculture. Removal of pathogens from the sewage prior to its reuse is needed not only for public health but also for maximum fish yield in ponds. Die off of pathogenic bacteria is considered to be a complex phenomenon in waste stabilisation ponds. The published articles on this subject consider, detention time and temperature the most influential parameters, while other factors have also been reported (Fernandez *et al.*, 1992; El-Hamouri *et al.*, 1995; Falowfield *et al.*, 1996; Rangeby *et al.*, 1996). These include presence of predators, high pH value and solar radiation. The high pH value resulting from algae photosynthetic activity plays an important role in promoting faecal coliform die-off in the pond system (El-Hamouri *et a*l., 1995). In duckweed-based pond, reduction of light penetration into the water column and slightly alkaline pH seems to reduce effective removal of pathogens. Islam (Islam *et al.*, 1990) reported that *Lemna minor* might serve as an effective environmental reservoir for *Vibrio cholera*. Dewedar and Bahgat (1995), reported that no decline in the counts were recorded in case of faecal coliform present in a set of dialysis sacs suspended under the duckweed mat, whereas faecal coliform in dialysis sacs exposed to sun light showed a decline with a decay rate constant of 0.1768 h^{-1}. Van der Steen (Van der Steen *et al.*, 2000) reported that the, faecal coliform decay in the dark parts of stabilisation ponds under conditions of carbon and nutrients sufficiency was negligible.

Duckweed production rate and protein content

Duckweed production rate and its protein content is controlled by climatic, environmental and water quality parameters. The nitrogen content of wastewater, quantitatively and qualitatively affects growth rate and protein content of duckweed. The duckweed yield and its protein content do not depend on the COD, however the yield and protein content depend significantly on the ammonia concentration (Oron *et al.*, 1984; Oron and Willers, 1989). Culley and Epps (1973) reported that the protein content of duckweed samples grown on treated municipal wastewater and untreated septic tank effluent was higher than the samples collected from animal and human wastes contaminated lagoons. Characteristics and production rates of different species of duckweed as reported in literature are presented in Table 4.

Table 4: Dry matter production rate of different species of duckweed and its protein and P content in g/100 g of dry matter

Duckweed sp.	Production Kg/ha/d	Dry matter[*]	Protein	P	References
Limna minor	-	3-5	14-25.9	0.57-0.8	Culley and Epps., 1973
Spirodela oligorrhiza	-	3-5	19.1-29.5	0.72-1.07	Culley and Epps, 1973
Lemna gibba	100-150	5.3	30	-	Oron and Willers, 1989
Lemna gibba	80-105	4.7-5.2	24.4-48.1	-	Oron *et al.*, 1984
Lemna gibba	44.5	7.8-8.8	15.5-28	0.48-0.86	Alaerts *et al.*, 1996
Lemna gibba	-	-	31.8-47.1	-	Hamouda *et al.*, 1995
Lemna gibba	74-164	-	-	-	Van der Steen *et al.*, 1998
Lemna Sp.	101-445	4.5	17.8	0.74	Ennabili *et al.*, 1998

[*] dry matter in g/100 g of fresh biomass

In Egypt two distinct duckweed species, *Lemna gibba* and *Lemna minor* have been reported. The growth rate of the former exceeds that of the latter, and the biomass production of *Lemna gibba* under field conditions is generally higher (Rejmankova, 1975; Porath *et al.*, 1979). Variation in the amino acids between the two species is negligible (Porath *et al.*, 1979). The duckweed *Lemna gibba* has been tested as a biological ammonia stripper in a combined circulating aquaculture of fish (Porath and Pollock, 1982). These authors reported that circulation of fishpond effluent in a duckweed pond with 30 cm depth promotes the uptake of ammonia from the effluent. The data showed that removal rate of ammonia is around 2 g/m2/day. Alaerts *et al.* (1996) suggested that microbial hydrolysis of the more complex organic N and P into NH_4^+ and ortho-PO_4^{-3} is the limiting step for enhancing biomass production in duckweed sewage stabilisation ponds. These authors added it is important to provide adequate pre-treatment for sewage (like UASB) to release organically bound N and P. Oron *et al*, (1987) mentioned that organic carbon represented as COD affects the efficiency of duckweed production in ponds. Pre-treatment of wastewater in a settling cone for about 8 hours decreased the organic carbon load and enhanced the ammonia uptake (Oron *et al.*, 1987). Pre-treatment of domestic sewage in a UASB could be useful in the improvement of duckweed ponds efficiency.

2.3. Fish Production

At present worldwide attention has been drawn to the possible loss of food security by the year 2010 (FAO, 2000). Food and Agriculture Organisation is reviewing the present and future state of food production, consumption and demand. FAO stated that fisheries and especially aquaculture sector will contribute in food security to the growing world population to the year 2010 and beyond.

2.3.1. World Fish Production

Marine and freshwater aquaculture is a rapidly growing form of food production making a major contribution to meeting the growing world demand for fish and fishery products. As aquaculture becomes more significant, recycling of wasted organic matter in fish aquaculture represents an attractive challenge. The application of inorganic fertilisers in fish farms facilitates eutrophication processes in the pond. The avoidance of unnecessary and harmful algae blooms could be achieved by changing from inorganic fertilisers to organic manure and supplementary feed. The reuse of wastes (organic wastes, animal wastes, agriculture wastes and human wastes) in fish production could safeguard the environment from pollution and lead to a valuable marketable biomass at low costs.

Over the next two decades, aquaculture will contribute more to the global food fish supplies and will help further reducing the global poverty and food insecurity, according to the strategy that has been established by FAO of the UN. Aquaculture's contribution towards the global fisheries landings continues to grow and reached 31.3% in 1999. Aquaculture production generates more food protein than any other animal food producing sector (FAO 2001). The total aquaculture production in 1999 was about 42.77 million metric tons, valued at 53.56 billion US$. Over the last three decades the sector has expanded, diversified and advanced technologically. The contribution of aquaculture to trade, both locally and internationally, has also increased and its share in the generation of income and employment for national economic development has also increased globally. The FAO has long recognised the importance of aquaculture to achieve world food security. A large proportion of global fish production comes from small-scale producers in developing countries and low-income food deficit countries. So aquaculture significantly contributes to food security, poverty alleviation and social-well being especially of the rural and peri-urban poor in many countries.

Globally most part of total fish production (catch and aquaculture) is coming from Asia. India stands as the sixth largest producer of total fish production, amounting to 4.95 million tons from its marine and fresh water ecosystems, of which about 2.24 million tons of fish is harvested from its inland aquatic resources. The country occupies the second position in inland fish production in the world (Sinha, 1999). In India aquaculture not only increased the export earnings, but also improved domestic supply resulting in annual per capita consumption of 8 Kg as compared to 2.8 kg in 1974 (Sinha, 1999). In 1988, China became the third country in the world whose yearly fish production was over 10 million tons. From 1990 onwards, China's yearly fish production ranked the first in the world (Zhiwen, 1999). Fish production was 20 million tons in 1994 and increased to 25.2 million tons in 1995, contributing about 20% of the world total fish production. The aquaculture output was 5.32 million tons in

1988 and reached 13.5 million tons by the end of 1995, accounting for 53.7% of the total China fish output (Zhiwen, 1999). More than two-thirds of aquaculture output is coming from freshwater aquaculture (9.4 million tons) and three quarters of this output is generated in pond-based culture.

Africa is the poorest continent in aquaculture production. Huisman (1985) documented that the fish production in the African region during 1985 was around 11,550 metric tons, representing only 0.3% of the global production. This production covered only 0.3% of the demand i.e. 34 grams out of 10.5 kg per capita per year. Nowadays aquaculture expansion in Africa is fast growing, however a number of constraints impose it. One of the most important constraints towards aquaculture development is the limited availability and high costs of supplementary feeds. In order to achieve improvements in the sector, attention has to focus on the exploration of locally available, suitable and cheap supplemental feeds (Allela, 1985). Huisman (1985) stated that the increment of aquaculture productivity in stagnant ponds or any other natural or artificial ecosystem could be achieved by increasing nutrient availability through fertilisation and/or supplementary feeding. The limited water resources and low availability of fish feed ingredients, limit aquaculture development. The availability of feed ingredients for supplementary fish feed is generally limited because of the scarce agriculture production and/or strong competition with livestock production. The major identified priorities for aquaculture development in Africa, as documented by FAO (Coche *et al.*, 1994), are the improvement of supplementary feeding strategies, organic fertilisation strategies and the promotion of integrated livestock-fish farming systems.

2.3.2. Fisheries production in Egypt

In Egypt, fisheries and aquaculture is an important component of the agriculture sector and represents a significant source of animal protein. It represented 4% of agriculture production and 14% of the total livestock and poultry production by value in 1994 (Shehadeh and Feidi, 1996). Egypt has been the traditional leader in aquaculture production in the Near East region, with a production of 73,000 ton out of 200,000 ton annually (ElGamal, 2001). In Egypt aquaculture started with traditional extensive and semi-intensive aquaculture systems. Rapid development has occurred in recent years to decrease the gap between the supply and demand for fish. In 1995, estimated total fish production was 407,000 ton, of which 71% was derived from marine and brackish water fisheries, 14% from freshwater fisheries and 15% from aquaculture (Shehadeh and Feidi, 1996). The production of cultured fish reached more than 19% of the total fish production in Egypt in 1997. The major part of Egyptian fish production (capture and aquaculture) is coming from inland fish production, where Egypt is ranked as 7th of the top ten countries in inland fish production (Table 5). Aquaculture production in brackish water is the main source and only 453 ton was harvested from marine aquaculture. The majority of fish farms in Egypt are semi-intensive brackish water farms. These farms are vulnerable because of the competition for land and water between aquaculture and agriculture. Scarcity of freshwater and land make freshwater aquaculture development restricted and limited by water and land policies. Now the main source of water for aquaculture is coming from agricultural drainage water. Intensive culture in earthen ponds and tanks is now developing quickly as a response to the potential drop in number of the semi-intensive farming systems. The national development strategy aims to increase annual per capita fish consumption from the present 10 kg to 13 kg by year 2017.

Table 5: Top ten countries in inland fish production (FAO, 2000)

Country	Production in 1998 (1000 tonnes)	% of world production (65% for top ten countries)
China	2280	28.5
India	650	8.1
Bangladesh	538	6.7
Indonesia	315	3.9
Tanzania	300	3.7
Russia	271	3.4
Egypt	253	3.2
Uganda	220	2.8
Thailand	191	2.4
Brazil	180	2.3

Extensive aquaculture is applied in inland lakes, rivers and irrigation canals. Brackish water lakes (e.g. in the Rayaan depression, 16,000 ha) are stocked with mullet and carp. The irrigation networks of the Nile valley have been stocked with grass carp for weed control since 1994, while black carp has been stocked since 1997. The harvest of grass carp amounted to 12,000 tons (ElGamal, 2001). In Egypt, semi-intensive aquaculture provides about 75% of Egypt's total aquaculture production and most farms are located in the North Eastern parts of the Nile delta and currently cover about 75,000 ha. The total area under pond culture was 67,300 ha, 89% of which are private farms and 11% governmental farms (Shehadeh and Feidi, 1996). Intensive aquaculture is not common in Egypt; only five farms are using intensive farming techniques and produce 500 ton annually, mostly tilapia. Integration of fish aquaculture with rice production is the common type of integrated agriculture that is practised in Egypt. Fish production from rice-fish integrated farms, peaked at 28,000 ton in 1989 and reduced to 7,000 ton in 1997. In 1995 rice-fish culture represented 172,800 ha while cage culture was 198,000 cubic meters (Shehadeh and Feidi, 1996). The species that are mostly farmed are tilapia (44% of the annual harvest) followed by mullets (25%), carps (24%) and sea bass and sea bream (3.5%). Most of aquaculture production, especially tilapia, carp and mullet, is consumed locally while marine species are produced for export.

2.3.3. Aquaculture feeds and pond fertilisers

Due to fast growing fish aquaculture now and in the future, there is growing demand on fish feed and pond fertilisers. The growing demand on pond fertilisers and fish feed makes it urgent for researchers and farmers to find new and different sources of feed. These sources are mostly depend on, the cost of source, culture technique and environmental regulation.

2.3.3.1. Use of animal manure, agricultural wastes and wild plants

Animal and agriculture wastes are being used to fertilise fish farms in many countries of the world. These wastes are considered to be superior to inorganic fertilisers in producing and maintaining desirable species of planktonic and benthic food organisms in fresh and brackish water ponds. Ecological agriculture or so called organic or biodynamic agriculture is different from chemical agriculture. The increase in agricultural production in case of chemical agriculture is due to high-energy input in the form of application of large quantities of synthetic fertilisers and pesticides. This in turn has caused a deterioration of farmlands, shortage of energy and resources, pollution of the environment and an upset of the ecological balance. Ecological agriculture is based on increment of both of utilisation efficiency of solar energy and conversion efficiency of biological energy, to enhance the circulation of energy and mass flow in the ecosystem. This type of agriculture saves the natural resources and keeps the environment free of pollution. The same principal can be applied in aquaculture.

In integrated livestock-cum-fish farming, the animals are raised near or on fishponds so that the manure and other waste materials can be discharged directly into the ponds. The solids settle to the pond bottom and the easily decomposable constituents undergo anaerobic decomposition, producing carbon dioxide, ammonia and inorganic phosphates. These are utilised by algae to grow and reproduce. The algal biomass produces oxygen by photosynthesis. This oxygen, along with that dissolved in the water from the atmosphere enhances the aerobic decomposition of the organic wastes added to the system. Studies (Pillay, 1992) confirmed that the non-mineralised fraction of the manure is used as food base by bacteria and protozoa. The competitive price of organic fertilisers, make it better than the inorganic fertilisers. Also organic wastes are considered to be superior to inorganic fertilisers in producing and maintaining desirable species of planktonic and benthic food organisms in fresh and brackish water ponds. Green et al. (1989) reported better fish yields in ponds supplied with chicken manure than in ponds fed with chemical fertilisers. On the other hand, use of chicken manure was similar to inorganic fertilisers, in fertilisation of ponds stocked with snakehead, milkfish and Nile tilapia and feed on supplementary feed (Cruz and Laudencia, 1980; Milstein et al., 1995). Fertilisation of ponds supplemented with supplementary feed resulted in higher fish yields (Cruz and Laudencia, 1980; Tidwell et al., 1995). The use of organic fertilisers with supplementary feed could meet the nutritional value of feed pellets. Tidwell et al. (1995) reported that the results for freshwater prawn showed no significant difference between the growth rates in ponds fed with formulated diet and those fed with supplementary feed in combination with pond fertilisation (Tidwell et al., 1995).

The integration of agriculture with aquaculture has received increasing attention in recent years with emphasis on the use of animal manure (chicken, geese, duck, sheep, and cattle) as nutrients in fishponds. Investigations were carried out on the integration of animal-fish-vegetable production. Sheep manure was used as nutrient in addition to formulated feed in a fish-cum-vegetable integrated production experiment using polyculture technique (Prinsloo and Schoonbee, 1987a). European common carp (Cyprinus carpio), Chinese silver carp (Hypophthalmichthys molitrix), Chinese grass carp (Ctenopharyngodon idella) and Chinese bighead carp (Aristichthys nobilis) were reared in ponds fed with sheep manure and pellet feed containing 18% protein. Effluent from the fishpond was used in vegetable production and compared with crop yield irrigated with fresh water enriched with inorganic fertiliser (Table 6).

Integrated culture systems combining fisheries, agriculture, animal husbandry and sideline crop production increase the utilisation rate of water bodies, land areas and the degree of self-sufficiency in fish feed supply. A more modern approach to this system is in use in India. Integration of raising poultry, ducks, cattle, sheep or pigs and growing vegetation such as mulberry, banana and other fruit trees on the fishpond embankment or adjacent to the ponds is a favourable approach. After introducing integrated composite fish culture technology in Indian villages, the poverty condition went down to 45.5% in 1990 from 89% in 1979 (Sinha, 1999). In order to enhance the natural productivity, pond water was fertilised with organic and inorganic fertilisers or wastewater was fed to the pond where the fish is normally fed with rice bran and oil cake.

Table 6: Vegetable production in ton/ha using effluent from sheep manure-fed fishponds and inorganic fertiliser enriched freshwater (Prinsloo and Schoobee, 1987a)

Vegetables species	Beetroot	Lettuce	Potatoes	Spinach	Tomatoes
Fish effluent	34.1	65.4	46.7	45.2	88.2
Freshwater	32.8	52.8	41.9	41.2	64.5

One component of the integrated waste-recycling system involves the use of filter-feeding fish to graze manure-fertilised plankton. Silver carp, bighead carp and tilapia were reared in ponds fertilised with swine manure (Behrends *et al.*, 1980). The production potential of duck-fish-vegetable integrated aquaculture-agriculture farming systems was carried out in developing areas in South Africa (Prinsloo and Schoonbee, 1987b). In Egypt, economic evaluation of, raising duck in tilapia-carp polyculture ponds in comparison with a non-integrated system (Soliman *et al.*, 2000) shows better net benefits in an integrated duck-fish system.

Table 7: Production and feed conversion ratio (FCR) for carps (grass carp and silver carp) fed two alternative feeds (Prinsloo and Schoobee, 1987c).

Items	Protein content[*]	Moisture content[**]	Feeding rate (Kg/ha/d)[***]	FCR[***]	Net fish yield (ton/ha/y)
Cabbage leaves	11.7	86.4	112	4.2	9.7
Kikuyu grass	8.7	73.6	163	6.3	9.5

[*] g/100 g of dry feed, [**] g/100g of fresh feed, [***] Dry matter bases

In recent years, great attention has been paid to aquatic plants and wild grasses as plant protein source for fish culture. Three kinds of aquatic plants (*Hydrilla sp., Lemna sp.,* and *Chara sp.*) were evaluated for raising Nile tilapia in cage culture (Rifai, 1980). In Bangladesh it was noted that feed containing duckweed is more effective than feed containing water hyacinth or rice bran in the diet of *Oreochromis niloticus* (Zaher *et al.*, 1995). The use of fresh duckweed as supplementary feed in carp polyculture in Bangladesh resulted in significantly higher yields than obtained with rice bran and oil cake (Azim and Wahab, 2003). In subtropical Nepal, the use of Napier grasses (*Pennisetum pupureum*) as the sole pond input in carp polyculture resulted in a net fish yield of 5.9 ton/ha/y. The use of fresh duckweed in raising tilapia and grass carp was performed in an intensive system as well as water static tanks (Gaigher *et al.*, 1984; Hassan and Edwards, 1992; Cui *et al.*, 1992; Shireman *et al.*, 1977). Prinsloo and Schoonbee (1987c) evaluate the use of cabbage leaves and Kikuyu grass as two alternative feeds in ponds stocked with grass carp (*Ctenopharyngodon idella*) and silver carp (*Hypophthalmichthys molitrix*). The fish production and feed conversion ratio is shown in Table 7. In China, aquatic grasses (*Vallisneria spiralis, Najas minor* and *Potamogeton malainus*) were used to feed polyculture Chinese carps and Wuchang fish stocked in pig manure fertilised ponds (Zhang *et al.*, 1987).

Table 8a: Prediction of global aquaculture production and feed requirements (Hasan, 2001) for 2010 and 2025

Fish species	Aquaculture production (1000 ton)		Feed requirement (1000 ton)	
	2010	2025	2010	2025
Carp	36,268	48,812	27,000	47,592
Tilapia	2,526	5,251	2,106	5,461
Shrimp & crabs	1,684	3,501	2,425	4,656
Salmon	1,569	3,760	1,255	2,632
Marine fish[*]	2,044	4,249	2,255	4,384
Trout	733	1,524	586	1,067
Catfish	604	1,256	761	1,551
Milkfish	462	536	554	638
Eel	263	305	284	290
Total	46,153	69,194	37,226	68,271

[*] Sea bass, sea bream, yellow tail, grouper, jacks and mullets, Flatfish, turbot, halibut, sole, cool and hake

Table 8b: Prediction of global use of fishmeal and fish oil (1000 ton) in aquaculture for 2010 and 2025 (Hasan, 2001)

Species	Percentage of fishmeal in the feed		Fishmeal Requirement		Percentage of fish oil in the feed		Fish oil requirement	
	2010	2025	2010	2025	2010	2025	2010	2025
Carp	2.5	0	675	0	0.5	0.5	135	238
Tilapia	3.5	2.5	74	136	0.5	0.5	11	27
Shrimp	20	15	485	698	3	3	73	14
Salmon	30	25	377	658	20	15	251	395
Marine fish[*]	40-45	30-40	931	1,413	12-15	10	321	438
Trout	25	20	147	213	20	15	117	160
Catfish	-	-	-	-	1	-	8	-
Milkfish	5	5	28	32	2	2	11	13
Eel	40	35	114	102	10	8	28	23
Total	-	-	2,831	3,262	-	-	955	1,308

[*] Sea bass, sea bream, yellow tail, grouper, jacks and mullets, Flatfish, turbot, halibut, sole, cool and hake

2.3.3.2. Use of formulated feed

It is well known that formulated feed is the most important cost factor in fish aquaculture. The predicted aquaculture development (Table 8) in the coming decades makes it urgent to partially replace fishmeal and high-price plant protein sources, with locally available low-cost ingredients. One of the requirements for reducing feed cost is incorporation of cheaper ingredients in the feed, like organic wastes and agro industrial by-products. Use of aquatic plants and low price plant protein is one of the current focal points in fish feed aquaculture research. The use of Azolla as substitute of fishmeal in the diets of Nile tilapia and medium carp was carried out (Basudha and Vishwanath, 1997; El-Sayed, 1992). Substitute of soybean by Azolla in the diet of *Clarias gariepinus* was studied by Fasakin and Balogun (1998). Substitution of plant protein with cocoa husk resulted in poor results with Nile tilapia (Falaye and Jauncey, 1999; Falaye *et al.*, 1999). El-Sayed (2003) investigated effects of different fermentation methods on the nutritive value of water hyacinth for Nile tilapia. The results showed that fermentation might only be necessary when incorporated into the feed pellets at ≥ 20% inclusion levels as substitute of wheat bran. Singh *et al* (2003) found that water socking of oil cakes for 24 hours before incorporation in the diet helped in the reduction of anti-nutritional factors. The replacement of fishmeal with salicornia meal in the feed of Nile tilapia fingerlings was performed and the results showed no significant difference up to 40% (Belal and Al-Dosari, 1999). The published results showed acceptance of soybean protein as replacement of fishmeal in the diets of red drum (Davis *et al.*, 1995) and catfish (Robinson and Li, 1994). The replacement of fishmeal with duckweed in the diets of tilapia showed no significance difference in the growth performance and nutrient utilisation up to 20% inclusion (Fasakin *et al.*, 1999). Virk and Saxena (2003) investigated good quality protein of amaranthus seeds that could be used as substitutes of rice bran and groundnut oil cake in the diets of common carp and rohu up to 50% inclusion levels. Feasibility of using poultry silage and culled sweet potatoes as feed ingredients for tilapia up to 33% was documented (Middleton *et al.*, 2001). Even some non-conventional animal protein sources like silkworm pupae were evaluated as replacement of fishmeal in carp feed (Rangacharyulu *et al.*, 2003).

2.3.3.3. Feed digestibility test in aquaculture

Partial or complete replacement of fishmeal with plant protein sources could be of considerable economic advantages even if this approach was associated with moderate reduction in the feed efficiency. Since feed formulation should be based on nutrient availability, reliable data on the digestibility of different ingredients are needed. Digestibility measurement is being used as the best way to evaluate nutritional value of diets and ingredients in fish aquaculture (Hajen *et al.*, 1993; Sullivan and Reigh, 1995; McGoogan and Reigh, 1996; Hasan *et al.*, 1997; Hossain *et al.*, 1997; Jones and De Silva, 1997; Martinez-Palacois, 1998). Difficulties to determine total faeces output in an aquatic system make it necessary to use indirect marker techniques for that purpose. The type of marker and faeces collection method significantly influence the digestibility coefficients (Van Den Berg and De La Noue, 2001). The most reliable and consistent estimates of nutrients digestibility were obtained by using chromic oxide and crude fibre as dietary markers while the performance of acid washed sand and polyethylene as dietary markers was disappointing (Tacon and Rodrigues, 1984). The values of digestibility were significantly lower with titanium dioxide as a marker than with chromic oxide in rainbow trout (Weatherup and McCraken, 1998). On the other hand, Sales and Britz (2001) reported that the Acid-Insoluble Ash (AIA) as an internal marker was the only marker that yielded consistent realistic apparent digestibility coefficients. The chromic oxide and crude fibre in faeces were either lower or similar to their respective levels in feed, resulting in negative apparent digestibility coefficient

(Sales and Britz, 2001). The naturally occurring AIA or AIA supplemented with Celite is good alternative marker for chromic oxide and more accurate and realistic than the acid-washed sand (Goddard and Mclean, 2001). Regarding faeces collection methods, Weatherup and McCraken (1998) reported that stripping and tank methods for faeces collection, both have inherent weaknesses, but the tank method gave accurate and replicable estimates of apparent digestibility of diets for fish. The objections to all the methods of direct faeces sampling (stripping) from the intestine relate to the fact that the faecal material may be removed prior to completion of natural retention time thereby reducing the digestion and absorption capacity and resulting in poor digestibility. Besides temperature (Kim et al., 1998), also the manufacturing process and treatment of feed (Gomes et al., 1993; Burel et al., 2000; Zhu et al., 2001) affect the digestibility coefficients.

The data of digestibility coefficients are available for some of the common ingredients. Amongst several plant-protein sources as partial substitutes of fishmeal, cottonseed meals, leguminous seeds, linseed, sesame and groundnut have been intensively tested as feed ingredients and data are available for different fish species (Reigh et al., 1990; Hossain et al., 1992; Nengas et al., 1995; Sullivan and Reigh, 1995; Fagbenro, 1996; Hasan et al., 1997; Fagbenro, 2001). There are no available data about the apparent digestibility coefficient of duckweed, fresh or dry, as fish feed ingredient for tilapia or any other fish species. Relatively few studies have been directed towards the assessment of duckweed as fish feed using herbivorous and omnivorous fish species. The experiments that dealt with the digestibility of aquatic plants and wild grasses were conducted using the endogenous markers. Estimation of dry matter and protein digestibility of duckweed species (*Lemna gibba* and *Lemna minor*) for grass carp was done by Van Dyke and Sutton, (1977) using natural internal markers. In 1979, Buddington calculated the digestibility coefficients of fresh *Najas guadalupensis* grass by *Tilapia zilli*. The data for fresh aquatic macrophyte *Hydrilla verticella* was reported for *Etroplus suratensis* (De Silva and Perera, 1983). These data for all these aquatic and wild plants have not been confirmed by using advanced indirect marker techniques. More attention for the apparent digestibility of aquatic and wild plants, especially duckweed, is needed.

2.3.3.4. Use of human sewage

Availability, cost and ease of transport are important considerations when selecting a nutrient source for fish cultivation. Chemical fertilisers, though widely used in the developed nations, are limited in supply and expensive in many developing countries. The cost of both chemical fertilizers and feed pellets is the major constraint to increase fish production in developing countries. For water, the second important element in fish aquaculture, assessment of global water scarcity predicts that by 2050 as much as 66% of the world population will experience some water stress and the water requirements for food production will be the major issue (Wallace and Gregory, 2002). The assessment expected reuse of treated wastewater in agricultural practices to emerge as one of the water resources, especially in water scarce countries. Use of domestic wastewater and human sewage for fish farming is an age-old practice in several Asian countries and is adopted in some others. Fish from sewage-fed ponds are an important source of animal protein for many millions of people in eastern Asia. The largest example of wastewater-fed aquaculture practices in the world is in Calcutta wetlands. There is 3000 ha of fishpond, which receives about 550,000 m^3/day of untreated wastewater and the ponds produce 13,000 ton of fish per year. In these ponds, carps are stocked at around 30,000 fingerlings per hectare with annual yield of 5-8 ton/ha in the better-managed ponds (Mara *et al.,* 1993; Kutty, 1980). In Indonesia fish are grown in bamboo cages in rivers that are heavily contaminated with human wastes in Bundung and Garut. In Thailand sludge from septic tanks is added to a single stage stabilisation-fishpond in which the nutrients in the waste are converted into plankton, which are harvested by Nile tilapia (Edwards, 1980). The Chinese also use sewage as a nutrient source for fish farming in semi closed-ecosystem (Wang, 1991). In this system, various integrated short food chains formed between fish, algae and aquatic macrophyte plants have made fish production rate as high as grain production rate, with much higher net income than the latter.

Aquaculture is applied for several purposes, culture for food production, improvement of natural stocks, sport fishing, recycling of organic wastes and production of industrial commodities. While all of these are of interest, the main interest in developing countries is the culture for food production and culture for recycling of organic wastes as it has the dual purpose of cleaning the environment and generating economic benefits. However, the use of untreated wastewater to feed fishponds has the disadvantage, that it may be difficult to ensure compliance with the current WHO guidelines for the microbiological quality of sewage-fed fishponds. In fact there are two problems relating to human consumption of fish raised on human wastes; the social acceptability of the fish and the potential for disease transmission. There appear to be definite cultural differences concerning the consumption of fish reared on human wastes. The Chinese and Indians appear to have few objections to eat such fish (Edwards, 1980). In the Arab world such suggestion will be rejected. There will be a big difference in acceptability if fish is directly fed on human wastes or fed on plankton grown in treated or partially treated sewage. Domestic sewage treatment in a UASB-duckweed ponds and subsequent reuse of the treated effluent and duckweed biomass in tilapia culture might be socially accepted in most cultural settings.

The cultivation of aquatic organisms (duckweed, algae, crustaceans and fish) has been a traditional way of improving water quality and facilitating nutrient recovery from wastewater in many parts of the world, especially developing countries. Efficiency of oxidation ponds for aerobic wastewater treatment could be enhanced via stocking the ponds with filter feeding fish. One of the major environmental problems in waste treatment in oxidation or high rate ponds is the algal bloom, which increases the biodegradable organic load in the effluent. Further, the nutrients concentrated by algae have to be removed or recovered from the effluent to prevent eutrofication of the water bodies and to meet the discharge standards set up by the regulatory agencies. Culture of herbivorous fish is an efficient way to reduce algae cells from the oxidation pond effluent. The biological processes involved are similar to those already described for animal and organic wastes.

Tilapia is fast growing and can tolerate low dissolved oxygen concentrations better than most fish. Adult tilapia is primarily herbivorous, occasionally omnivorous and some species are used to control aquatic weeds. In one experiment in Mirzapur, Bangladesh, in spite of the absence of tilapia in the polyculture group at the beginning of stocking it enters the ponds from the natural waterways and contributes to about 40% of the total fish production. Tilapia appeared to be able to harvest food from all of the space and food niches in the pond. Therefore it is tested in Mirzapur as an alternative to carp polyculture using duckweed as fish feed (Skillicorn *et al.*, 1993).

Preliminary experiments were conducted in a combined waste treatment-recycling system consisting of three 200 m^2 high rate stabilisation ponds, a series of 4 m^3 concrete fishponds, and a maize plot (Edwards and Sinchumpasak, 1981; Edwards *et al.*, 1981). The sewage was used to feed a shallow, high rate stabilisation pond in which the waste is rapidly treated and converted into algae biomass. The algae containing effluent is then pumped into a pond containing Nile tilapia. To fully recycle the nutrients contained in the waste, the effluent from the fishponds was used to fertilise maize.

As pathogens are the main concern of sewage reuse in fish aquaculture, Mara *et al* (1993) stated that use of a low cost treatment system comprising a 1 day anaerobic pond followed by a 5 days facultative pond prior to discharge into the fishpond could improve microbial quality of fish. Based on this design, fish culture experiments were conducted in Egypt using a laboratory scale model and a full-scale model (Easa *et al.*, 1995; El-Gohary *et al.*, 1995; Shereif and Mancy, 1995; Shereif *et al.*, 1995). In the laboratory scale model Nile tilapia grew healthy during the study period. In the full-scale model, treated sewage was successfully used to grow two types of local fish (*Oreochromis niloticus* and *Mugil sehli*). Fish production without supplemental feeding or pond aeration, reached 5-7 ton/ha/year. Furthermore, the fish were found to be free from human parasites and safe for human consumption. Nasr *et al* (1998) replaced the anaerobic pond with a UASB reactor and the facultative and maturation pond with an algae pond. The effluent from the UASB-algal pond system treating domestic sewage was used to fertilise fishpond stocked with Nile tilapia (Nasr *et al.*, 1998). The results revealed that the fish production in ponds receiving algal pond effluent was good provided that un-ionised ammonia was below 0.33 mg/l. In another experiment, primary and secondary treated sewage from an activated sludge system were used to fertilise an oxidation pond, in which Nile tilapia was reared (Khalil and Hussein, 1997). The microbiological analyses revealed that there was no evidence of any public health hazard associated with wastewater reuse in aquaculture.

Fish are known to harbour enterobacteriaceae that cause disease in human or other warm-blooded animals (Pillay, 1992). Vibrios causing epidemics such as cholera have been isolated from a population of healthy juvenile *Penaeus vannamei* (Gomez *et al.*, 1998). The sediment of the fishpond serves as a pool of indicator bacteria. The population of indicator bacteria together with some limnological parameters were monitored during a 6 years study of 3 eutrophic ponds (Markosova and Jezek, 1994). The results demonstrated that fish stocking could affect bacterial population in the sense that during high fish biomass the number of indicator bacteria as well as organic matter and phytoplankton were higher. Similar results were reported with high levels of faecal coliform in the effluent of flow-through commercial catfish ponds (Davis *et al.*, 1995).

Digestive tracts of fish harbour a wide spectrum of bacteria. In experimental studies faecal coliform, faecal streptococci were isolated from gut contents of fish grown in domestic wastewater-fertilised ponds and in some cases from the muscle tissues (Hejkal *et al.,* 1983). Digestive tracts of tilapia reared in commercial fishponds with animal manure or/and municipal wastewater generally were contaminated with faecal indicator bacteria to levels higher than those in the environment (Fattal *et al.,* 1993). When the densities of indicators in the water are sufficiently high, liver and muscles could also be contaminated. Cooking of fish prior to human consumption makes the major public health concern limited to the risk of Aeromonas wound infections among individuals who handle and process fish. Even shortly cooking of sewage-grown fish, by boiling in water or frying in oil makes fish free of pathogen (Kutty, 1980; Pillay, 1992; Fattal *et al.,* 1993).

2.3.4. Factors affecting growth performance and fish yield

Like any other ecosystem, fishponds are affected by a number of biotic and abiotic factors. Effective water quality management and fish production in fishponds can be achieved through better understanding of these factors. Feed management and reduction of aquaculture wastes through feeding strategies are very important not only to save feed costs but also to control water quality. Undigested (faecal) and uneaten feed in addition to the soluble end products of nitrogen and phosphorus metabolism can easily lead to environmental problems in the pond. High levels of nutrients from metabolic by-products, faecal materials and unconsumed feed cause algae blooms especially of toxic species that subsequently cause environmental hazards, including fish mortality (Pillay, 1992). Oxygen consumption by dense algae during the night period may result in low levels of oxygen concentration. In addition, the microbial degradation of large masses of algae can give rise to low levels of dissolved oxygen and promote growth of undesirable macrophytes. Optimal feeding regimes therefore are crucial in preventing the problems. In well-managed ponds wastage of feed is minimised by optimising the feeding rate and feeding regimes. Fish farmers try to adjust feed inputs in response to environmental variables. The bio-deposits utilise considerable quantities of oxygen in oxidising the organic matter contained in them, and eventually create a reducing environment that results in production of additional amounts of ammonia and hydrogen sulphide. The accumulation of faeces and wasted feed also cause an increase in the numbers of microorganisms, especially bacterial population (Pillay, 1992). All of these negatively affect oxygen concentration in the pond and has a detrimental effect on feed conversion ratio in all fish species (Prinsloo and Schoonbee, 1987c).

A main concern of fish farmers is management of risks associated with the oxygen budget in the pond. Exposure of fish to water hypoxia affects metabolic and physiologic processes (Schmitt and Uglow, 1998). Under such conditions, strong negative correlation between oxygen concentration and feed conversion ratio was confirmed (Thomas and Piedrahita, 1997). Water hypoxia has indirect detrimental effect by enhancing microbial transformation of nitrate into nitrite especially in the presence of sewage bacteria (Anuradha and Subburam, 1995). The nitrite has its own role in fish toxicity. It causes brown blood disease in fish and blue babies syndrome in infants. Its toxicity is attributed to oxidation of haemoglobin into non-functional methemoglobin. Colt *et al.* (1981) confirm negative effects of nitrite on the growth performances of channel catfish. Behrends *et al* (1980) also reported that stress associated with low dissolved oxygen concentration, may contribute to bacterial disease. The low dissolved oxygen levels increase nitrite toxicity and both low dissolved oxygen and nitrite toxicity increase susceptibility of the fish to bacterial infection (Bunch and Bejerano, 1997).

Un-ionised ammonia is the most toxic parameter, not only to fish but to all aquatic animals (Baird *et al.*, 1979; Zhao *et al.*, 1997; Harris *et al.*, 1998; Nasr *et al.*, 1998). The toxic levels of ammonia for short-term exposure usually lie between 0.6 and 2 mg/l, while others consider the maximum tolerable concentration to be 0.1 mg/l (Pillay, 1992). The usual ambient ammonia concentrations under intensive farming conditions might affect growth performance of fish. This has been supported by Person-Le Ruyet *et al* (1997) who reported that feed efficiency and growth performance are negatively affected by ambient ammonia concentration, due to major changes in food conversion ratio, and protein utilisation (Person-Le Ruyet *et al.*, 1997). Quality of feed and especially dietary protein is of fundamental importance to aquaculture systems, representing considerable economic investment and a central factor determining fish growth. If the maximum utilisation of dietary protein for growth is to be achieved, the selection for quantity and quality of protein is important. Ingested feed nitrogen in fish is either excreted as ammonia or stored in fish biomass. The dietary protein concentration determines how much of the ammonia is excreted by fish (Buttle *et al.*, 1995; Brunty *et al.*, 1997, Chakraborty and Chakraborty, 1998; Thomas and Piedrahita, 1998). This is the reason why farmers quantify feed and protein requirements. This reduces feed costs, maximises growth and maintains a good water quality.

There are two primary processes that affect ammonia concentration, ammonia excretion and sediment diffusion (Hargreaves, 1997). These two processes are sensitive to dietary protein digestibility, dietary protein level and phytoplankton specific uptake rate. In an integrated aquatic plants-fish aquaculture system ammonia concentration will increase, decrease, or remain stable over time if total ammonia production by fish is greater than, less than, or equal to ammonia assimilation by plants plus ammonia losses, respectively (Seawright *et al.*, 1998). In an intensive aquaculture system, the nitrogen dynamics in the water column is controlled by nitrification, which is the dominant process in the system. In ponds assimilation by phytoplankton and subsequent sedimentation is the principal process of ammonia removal (Lorenzen *et al.*, 1997). Gomez *et al.* (1995) reported that ammonia concentration is negatively correlated with pH, temperature and chlorophyll concentration and positively with total nitrogen input. In ideal fish ponds ammonia volatilisation is limited. Dissolved oxygen concentration has a key role in the promotion of nitrification as one of the processes for the removal of ammonia (Liang *et al.*, 1998). The process of nitrification is limited in ponds with more sediment (Diab *et al.*, 1993). Use of fresh duckweed might limit ammonia accumulation in ponds. The fresh duckweed remains floated on the pond surface and therefore can not be wasted. The photosynthetic active duckweed might even absorb some of the excreted ammonia as long as it remains uneaten on the surface.

Optimisation and better understanding and management of duckweed-based fish aquaculture systems to obtain high nutrient recovery using domestic sewage are the challenges of the current research.

3. Scope of this dissertation

The research work presented in this thesis focuses on the nutrients recovery and water recycling from domestic sewage in fish farming by using a UASB-duckweed ponds wastewater treatment system, integrated with fish farming ponds. Chapter 2 focuses on the nutrients recovery from domestic sewage by using a UASB-duckweed ponds system. The duckweed production, its quality and its nutrients uptake capacity under Egyptian climatic condition were assessed. The treatment efficiency and quality of the final effluent during warm and cold season were described. Special attention was paid to the nitrogen mass balance and effect of ambient temperature on the treatment efficiency.

In chapter 3, the evaluation of duckweed, fresh and dry, as sole fish feed for tilapia production in a semi-static water system was assessed under different feeding rates within a temperature range of 16-25 °C. The evaluation of fresh duckweed and wheat bran as supplementary fish feed and the evaluation of treated sewage and fresh water as water sources were conducted in chapter 4. A comparison between UASB-duckweed-fish farming system with conventional sewage-fed fish farm was made.

The effect of ammonia as one of the most important water quality parameters was investigated by studying the chronic toxicity of ammonia on fresh duckweed-fed tilapia (chapter 5). Microbial quality of tilapia (*Oreochromis niloticus*), reared in different faecal-contaminated fishponds, was examined in chapter 6. In chapter 7, fresh duckweed from the treatment system was compared with fishmeal-based commercial feed. Treated effluent and freshwater were also evaluated as two alternative water sources while applying these feeds. In chapter 8 measurements of apparent digestibility coefficients of fresh and dry duckweed and its use in intensive tilapia farming are reported. The main conclusions, discussion and recommendations are presented in chapter 9.

References

Abdalla, A.L., Ambrosano, E.J., Vitti, D.M.S.S. and Silva, F.J.C. 1987. Water hyacinth (Eichhornia *crassipes*) in ruminant nutrition. Water. Sci. Technol. 19(10), 109-112.

Abdel Mageed, Y. 1994. Water in the Arab world, perspectives and prognoses, the central region: problems and perspective. In: Peter Rogers and Peter Lydon (Eds.), Water in the Arab world, perspectives and prognoses, pp. 101-119

Alaerts, G.J., Mahbubar, M.D.R. and Kelderman, P. 1996. Performance analysis of a full-scale duckweed-covered sewage lagoon. Water Res. 30(4), 843-852.

Allela, S.D. 1985. Development and research of aquaculture in Kenya. Aquaculture Research in the African region "proceeding of the African Seminar on aquaculture, organized by the IFS, 7-11 October 1985, Pp. 112-114

Al-Nozaily, F., Alaerts, G. and Veenstra, S. 2000. Performance of duckweed-covered sewage lagoon. 2. Nitrogen and phosphorus balance and plant productivity. Water Res. 34(10), 2734-2741

Anuradha, S. and Subburam, V. 1995. Role of sewage bacteria in methemoglobin formation in *Cyprinus carpio* exposed to nitrate. J. Envi. Biol. 16(2), 175-179.

Azim, M.E. and Wahab, M.A. 2003. Development of a duckweed-fed carp polyculture system in Bangladesh. Aquaculture 218, 425-438

Baird, R., Bottomley, J. and Taitz, H. 1979. Ammonia toxicity and pH control in fish toxicity bioassays of treated wastewater. Water Res. 13, 181-184.

Barbosa, R.A. and Sant Anna, G.L. jr, 1989. Treatment of raw domestic sewage in an UASB reactor. Water Res. 23, 1483-1490

Basudha, C. and Vishwanath, W. 1997. Formulated feed based on aquatic weed Azolla and fishmeal for rearing medium carp *Osteobrama belangeri* (Valenciennes). J. Aquacult. Trop. 12, 155-164.

Behrends, L.L., Maddox, J.J., Madewell, C.E. and Pile, R.S. 1980. Comparison of two methods of using liquid swine manure as an organic fertilizer in the production of filter-feeding fish. Aquaculture 20, 147-153.

Belal, I.E.H. and Al-Dosari, M. 1999. Replacement of fishmeal with Saicornia meal in feeds for Nile tilapia (*Oreochromis niloticus*). Journal of the world aquaculture society 30(2), 285-289

Bunch, E.C. and Bejerano, I. 1997. The effect of environmental factors on the susceptibility of hybrid tilapia *Oreochromis niloticus* X *Oreochromis aureus* to streptococcus. Bamidgeh 49(2), 67-76.

Buddington, R.K. 1979. Digestion of an aquatic macrophyte by *Tilapia zillii* (Gervais) J. Fish Biol. 15, 449-455

Burel, C., Boujard, T., Tulli, F., Kaushik, S.J. 2000. Digestibility of extruded peas, extruded lupin, and rapeseed meal in rainbow trout (*Oncorhynchus mykiss*) and turbot (*Psetta maxima*). Aquaculture 188, 285-298

Buttle, L.G., Uglow, R.F. and Cowx, I.G. 1995. Effect of dietary protein on the nitrogen excretion and growth of the African catfish, *Clarias gariepinus*. Aquat. Living Resour. 8, 407-414.

Brunty, J.L., Bucklin, R.A., Davis, J., Baird, C.D. and Noedstedt, R.A. 1997. The influence of feed protein intake on tilapia ammonia production. Aquacult. Eng. 16, 161-166.

Chakraborty, S.C. and Chakraborty, S. 1998. Effect of dietary protein level on excretion of ammonia in Indian major carp, *Labeo rohita*, fingerlings. Aquacult. Nutr. 4, 47-51.

Coche, A.G., Haight, B.A. and Vincke, M.M.J. 1994. Aquaculture development and research in sub-Saharan Africa "CIFA technical paper, FAO, Rome.

Colt, J., Ludwig, R., Tchobalongolous, G. and Cech, J. 1981. The effects of nitrite on the short-term growth and survival of channel catfish, *Ictalurus punctatus*. Aquaculture 24, 111-122.

Cruz, E.M. and Laundencia, I.L. 1980. Poluculture of milkfish (*Chano chanos* Forskal), all-male Nile tilapia (*Tilapia nilotica*) and snake head (*Ophicephalus striatus*) in freshwater ponds with supplemental feeding. Aquaculture 20, 231-237.

Cui, Y., Liu, X, wang, S. and Chen, S. 1992. Growth and energy budget in young grass carp, *ctenopharyngodon idella* val., fed plant and animal diets. J. Fish Biol. 42, 231-238

Culley, Jr D.D. and Epps, E.A. 1973. Use of duckweed for waste treatment and animal feed. J. W. P. C. F. 45(2), 337-347.

Davis, E. M., Mathewson, J. J. and de la Cruz, A. T. 1995. Growth of indicator bacteria in a flow-through aquaculture facility. Water Res. 29(ii), 2591-2593.

De Silva, S.S. and Perera, M.K. 1983. Digestibility of an aquatic macrophyte by the Cichlid Etroplus suratensis (Bloch) with observations on the relative merits of three indigenous components as markers and daily changes in protein digestibility. J. Fish. Biol. 23, 675-684

Dewedar, A. and Bahgat, M. 1995. Fate of faecal coliform bacteria in a wastewater retention reservoir containing *Lemna gibba* L. Water Res. 29(11), 2598-2600.

Diab, S., Kochab, M. and Avnimelech, Y. 1993. Nitrification pattern in a fluctuating anaerobic-aerobic pond environment. Water Res. 27(9), 1469-1475.

Dixo, N.G.H., Gambrill, M.P., Catunda, P.F.C. and van Haandel, A.C. 1995. Removal of pathogenic organisms from the effluent of an up-flow anaerobic digester using waste stabilization ponds. Water Sci. Technol. 31 (12), 275-284.

Draaijer, H., Maas, J.A.W., Schaapman, J.E. and Khan, A. 1992. Performance of the 5MLD UASB reactor for sewage treatment at Kanpur, India. Water Sci. Technol. 25(7), 123-133.

Easa, M.E., Shereif, M.M., Shaaban, A.I. and Mancy, K.H. 1995. Public health implication of wastewater reuse for fish production. Water Sci. Technol. 32 (11), 145-152.

Edwards, P. 1980. A review of recycling organic wastes into fish, with emphasis on the tropics. Aquaculture 21, 261-279.

Edwards, P. and Sinchumpasak, O. 1981. The harvest of microalgae from the effluent of sewage fed high rate stabilization pond by *Tilapia nilotica* part 1: description of the system and the study of the high rate pond. Aquaculture 23, 83-105.

Edwards, P., Sinchumpasak, O. and Tabucanon, M. 1981. The harvest of microalgae from the effluent of sewage fed high rate stabilization pond by *Tilapia nilotica* part2: studies of the fish ponds. Aquaculture 23, 107-147.

Edwards, P., Kaewpaitoon, K., Little, D.C. and Siripandh, N. 1994. An assessment of the role of buffalo manure for pond culture of tilapia. II: field trial. Aquaculture 126, 97-106.

ElGamal, A.R. 2001. Status and development of aquaculture in the Near East. In R.P. Subasinghe, P. Bueno, M.J. Phillips, G. Hough, S.E. McGladdery and J.R. Arthur, eds. "Aquaculture in the third millennium", technical proceedings of the conference on aquaculture in the third millennium, Bangkok, Thailand, 20-25 February 2000, Pp. 357-376, NACA, Bangkok and FAO, Rome.

El-Gohary, F., El-Hawarry, S., Badr, S. and Rashed,.Y. 1995. Wastewater treatment and reuse for fish aquaculture. Water Sci. Technol. 32 (11), 127-136.

El-Gohary, F. 2002. Need for alternative water resources in Egypt. Use of appropriately treated wastewater in irrigated agriculture, technical and socio-economic aspects. Workshop organized in the framework of multilateral research and capacity building projects in the Middle East/Mediterranean area, Wageningen University, sub-department of environmental technology and, Irrigation and water engineering group, April 24th 2002, Wageningen, The Netherlands.

El-Hamouri, B., Jellal, J., Outabiht, H., Nebri, B., Khallayoune, K., Benkerroum, A., Hajli, A. and Firadi, R., 1995. The performance of a high-rate algal pond in the Moroccan climate. Water. Sci. Tech. 31(12), 67-74

El-Sayed, A.F.M. 1992. Effects of substituting fish meal with Azolla pinnata in practical diets for fingerling and adult Nile tilapia, oreochromis niloticus L. Aquaculture and Fisheries Management 23, 167-173

El-Sayed, A.M. 2003. Effects of fermentation methods on the nutritive value of water hyacinth for Nile tilapia *Oreochromis niloticus* (L.) fingerlings. Aquaculture 218, 471-478

El-Sharkawi, F., El-Sebaie, O., Hossam, A. and Abdel Kerim, G. 1995. Evaluation of Daqahla wastewater treatment, aerated lagoon and ponds system. Water Sci. Technol. 32(11), 111-119.

Ennabili, A., Ater, M. and Radoux, M. 1998. Biomass production and NPK retention in macrophytes from wetlands of the Tingitan peninsula. Aquatic Botany 62, 45-56.

Fagbenro, O.A. 1996. Apparent digestibility of crude protein and gross energy in some plant and animal-based feedstuffs by *Clarias isheriensis* (Siluriformes: Clariidae) (Sydenham 1980). J. Appl. Ichthyol. 12, 67-68

Fagbenro, O.A. 2001. Apparent digestibility of crude protein and gross energy in some plant and animal-based feedstuffs by *Heterotis niloticus* (Dupeiformes: Osteoglossidae) (Luvier 1829). J. Aquault. Trop. 16(3), 277-282

Falaye, A.E. and Jauncey, K.C. 1999. Acceptability and digestibility by tilapia *Oreochromis niloticus* of feed containing cocoa husk. Aquacult. Nutr. 5, 157-161

Falaye, A.E., Jauncey, K. and Tewe, O.O. 1999. The growth performance of tilapia (Oreochromis niloticus) fingerlings fed varying levels of cocoa husk diets. J. Aquault. Trop. 14(1), 1-10

Fallowfield, H.J., Cromar, N.J. and Evison, L.M. 1996. Coliform die-off rate constants in a high-rate algal pond and the effect of operational and environmental variables. Water Sci. Technol. 34(11), 141-147

FAO 2000. The state of the world fisheries and aquaculture, edited by editorial group, FAO information division.

FAO 2001. Aquaculture in the third millennium, technical proceedings of the conference on aquaculture in the third millennium, edited by Subasinghe, R.P., Bueno, P., Phillips, M.J., Hough, G., McGladdery S.E. and Arthur, J.R., Bangkok, Thailand, 20-25 February 2000, NACA, Bangkok and FAO, Rome.

Fasakin, A.E. and Balogun, A.M. 1998. Evaluation of dried water fern (*Azolla pinnata*) as a replacer for soybean dietary components for Clarias gariepinus fingerlings. J. Aquault. Trop. 13(1), 57-64

Fasakin, E.A., Balogun, A.M. and Fasuru, B.E. 1999. Use of duckweed, Spirodela polyrrhiza L. schleiden, as a protein feed stuff in practical diets for tilapia, oreochromis niloticus L. Aquacult. Res. 30, 313-318

Fattal, B., Dotan, A., Parpari, L., Tchorsh, Y. and Cabelli, V.J. 1993. Microbiological purification of fish grown in faecally contaminated commercial fishpond. Water Sci. Technol. 27(7-8), 303-311.

Ferguson, A.R. and Bollard, E.G. 1969. Nitrogen metabolism of Spirodela oligorrhiza I-Utilization of ammonia, nitrate and nitrite. Planta (Berl) 88, 344-352.

Fernandez, A., Tejedor, C. and Chodi, A. 1992. Effect of different factors on the die-off of faecal bacteria in a stabilization pond purification plant. Water Res. 26(8), 1093-1098

Gaigher, I.G., Porath, D. and Granoth, G. 1984. Evaluation of duckweed (*Lemna gibba*) as feed for tilapia (*Oreochromis niloticus × Oreochromis aureus*) in a recirculating unit. Aquaculture 41, 235-244.

Goddard, J.S. and Mclean, E. 2001. Acid insoluble ash as an inert reference material for digestibility studies in tilapia, *Oreochromis aureus*. Aquaculture 194, 93-98

Gomes, E.F., Corraz, G. and Kaushik, S. 1993. Effects of dietary incorporation of co-extruded plant protein (rapeseed and peas) on growth, nutrient utilization and muscle fatty acid composition of rainbow trout (*Oncorhynchus mykiss*). Aquaculture 113, 339-353

Gomez E., Casellas C., Picot B. and Bontoux J. (1995) Ammonia elimination processes in stabilization and high-rate algal pond systems. Water Sci. Technol. 31(12), 303-312.

Gijzen, H.J. 2001. Anaerobes, aerobes and phototrophs: a wining team for wastewater management. Water Sci. Technol. 44(8), 123-132

Gijzen, H.J. 2002. Anaerobic digestion for sustainable development: a natural approach. Water Sci. Technol. 45(10), 321-328

Gomez, G.B., Tron, M.L., Ana Roque, Turbull, J.F., Inglis, V. and Guerra, F.A. 1998. Species of Vibrio isolated from hepatopancreas, haemolymph and digestive tract of a population of healthy juvenile *Penaeus vannamei*. Aquaculture 163, 1-9.

Green, B.W., Pheleps, R.P. and Alvarenga, H.R. 1989. The effects of manures and chemical fertilizers on the production of *Oreochromis niloticus* in earthen ponds. Aquaculture 76, 37-42.

Hajen, W.E., Higgs, D.A., Beames, R.M. and Dosanjh, B.S. 1993. Digestibility of various feed stuffs by post-juvenile chinook salmon (*Oncorhynhus tshawytscha*) in seawater: 2-Measurment of digestibility. Aquaculture 112, 333-348

Hammouda, O., Gaber, A. and Abdel-Hameed, M.S. 1995. Assessment of the effectiveness of treatment of wastewater-contaminated aquatic systems with *Lemna gibba*. Enz. and Micro. Technol. 17, 317-323.

Hargreaves, J. A. 1997. A simulation model of ammonia dynamics in commercial catfish ponds in the southeast United States. Aquacult. Eng. 16, 27-43.

Harris, J.O., Maguire, G.B., Edwards, S. and Hindrum, S.M. 1998. Effect of ammonia on the growth rate and oxygen consumption of juvenile greenlip abalone, *Haliotis laevigata* Donovan. Aquaculture 160, 259-272.

Harvey, R. M. and Fox, J. L. 1973. Nutrient removal using *Lemna minor*. J. W. P. C. F. 45(9), 1928-1938.

Hassan, M.S. and Edwards, P. 1992. Evaluation of duckweed (Lemna perpusilla and Spirodela polyrrhiza) as feed for Nile tilapia (*Oreochromis niloticus*). Aquaculture 104, 315-326.

Hasan, M.R., MaCintosh, D.J. and Jauncey, K. 1997. Evaluation of some plant ingredients as dietary protein sources for common carp (*Cyprinus carpio* L.) fry. Aquaculture 151, 55-70

Hasan, M.R. 2001. Nutrition and feeding for sustainable aquaculture development in the third millennium. In R.P. Subasinghe, P. Bueno, M.J. Phillips, G. Hough, S.E. McGladdery and J.R. Arthur, eds. "Aquaculture in the third millennium", technical proceeding of the conference on aquaculture in the third millennium, Bangkok, Thailand, 20-25 February 2000, Pp. 193-219, NACA, Bangkok and FAO, Rome.

Hejkal, T.W., Gerba, C.P., Henderson, S. and Freeze, M. 1983. Bacteriological, virological and chemical evaluation of a wastewater-aquaculture system. Water Res. 17(12), 1749-1755.

Hossain, M.A., Nahar, N., Kamal, M. and Islam, M.N. 1992. Nutrient digestibility coefficients of some plant and animal proteins for Tilapia (*Oreochromis niloticus*). J. Aquacult. Trop. 7, 257-266

Hossain, M.A., Nahar, N. and Kamal, M. 1997. Nutrient digestibility coefficients of some plant and animal proteins for rohu (*Labeo rohita*). Aquaculture 151, 37-45

Huisman, E.A. 1985. Current status and role of aquaculture with special reference to the African region. Aquaculture Research in the African region "proceeding of the African Seminar on aquaculture, organized by the IFS, 7-11 October 1985, Pp. 11-22

Islam, S.I., Drasar, B.S. and Bradley, N.J. 1990. Survival of toxigenic Vibrio cholerae 01 with common duckweed, Lemna minor in artificial aquatic ecosystems. Transaction of the Royal Society of Tropical Medicine and Hygiene 84, 422-424.

Jones, P.L. and De Silva, S.S. 1997. Apparent nutrient digestibility of formulated diets by the Australian freshwater crayfish *Cherax destructor* Dark (Decapoda, Parastacidae). Aquacult. Res. 28, 881-891

Kawai, H., Uehara, M.Y., Gomes, J.A., Jahnel, M.C., Rossetto, R., Alem, S.P., Ribeiro, M.D., Tinel, P.R. and Greico, V.M. 1987. Pilot-scale experiments in water hyacinth lagoons for wastewater treatment. Water Sci. Technol. 19(10), 129-173.

Kell, A.D.K., Saada, T.A. and Somerville, J.P. 1993. Greater Cairo Wastewater Project, Proceedings of the institution of civil engineers, institution of civil engineers, one Great George Street, London SW1P3AA, pp. 8-17.

Khalil, M.T. and Hussein, H.A. 1997. Use of wastewater for aquaculture: an experimental field study at a sewage treatment plant, Egypt. Aquacult. Res. 28, 859-865.

Kim, J.D., Breque, J. and Kaushik, S.J. 1998. Apparent digestibility of feed components from fishmeal or plant protein based diets in common carp as affected by water temperature. Aquat. Living Resour. 11(4), 269-272

Körner, S. and Vermaat, I.E. 1998. The relative importance of *Lemna gibba* L., bacteria and algae for nitrogen and phosphorous removal in duckweed-covered domestic wastewater. Water Res. 33(12), 3651-3661.

Kutty, M.N. 1980. Aquaculture in South East Asia: some points of emphasis. Aquaculture 20, 159-168.

Liang, Y., Cheung, R.Y.A., Everitt, S. and Wong, M.H. 1998. Reclamation of wastewater for polyculture of freshwater fish: wastewater treatment in ponds. Water Res. 32(6), 1864-1880.

Lorenzen, K., Struve, J. and Cowan, V.J. 1997. Impact of farming intensity and water management on nitrogen dynamics in intensive pond culture: a mathematical model applied to Thai commercial shrimp farms. Aquacult. Res. 28, 493-507.

Mara, D.D., Edwards, P., Clark, D. and Mills, S.W. 1993. A rational approach to the design of wastewater-fed fish ponds. Water Res. 27(12), 1797-1799.

Markosova, R. and Jezek, J. 1994. Indicator bacteria and limnological parameters in fish ponds. Water Res. 28(12), 2477-2485.

Martinez-Palacios, C.A. 1988. Digestibility studies in juveniles of the Mexican cichlid, *Cichlasoma urophthalmus* (Gunther). Aquaculture and Fisheries Management 19, 347-354

Mbagwu, I.G. and Adeniji, H.A. 1988. The nutritional content of duckweed (*Lemna paucicostata* Hegelm) in the Kainji Lake area, Nigeria. Aquat. Bot. 29, 357-366.

McGoogan, B.B. and Reigh, R.C. 1996. Apparent digestibility of selected ingredients in red drum (*Sciaenops ocellatus*) diets. Aquaculture 141, 233-244

Mergaert, K., van der Haegen, B. and Verstraete, W. 1992. Applicability and trends of anaerobic pre-treatment of municipal wastewater. Water Res. 26 (8), 1025-1033.

Middleton, T.F., Fekert, P.R., Boyd, L.C., Daniels, H.V. and Gallagher, M.L. 2001. An evaluation of co-extruded poultry silage and culled jewel sweet potatoes as a feed ingredient for hybrid tilapia (*Oreochromis niloticus × Oreochromis mossambicus*). Aquaculture 198, 269-280

Milstein, A., Alkon, A. and Karplus, I. 1995. Combined effects of fertilization rate, manuring and feed pellet application on fish performance and water quality in polyculture ponds. Aquacult. Res. 26, 55-65

Mokady, S., Yannai, S., Einav, P. and Berk, Z. 1979. Algae grown on wastewater as a source of protein for young chicken and rats. Nutr. Rep. Inter. 19(3), 383-390.

Morris, P.F. and Barker, W.G. 1977. Oxygen transport rates through mats of *Lemna minor* and *Wolffia sp.* and oxygen tension within and below the mat. Can. J. Bot. 55, 1926-1932.

Nasr, F.A., El-Shafai, S.A. and Abo-Hegab, S. 1998. Suitability of treated domestic wastewater for raising *Oreochromis niloticus*. Egypt. J. Zool. 31, 81-94

Nelson, S.G., Smith, B.D. and Best, B.R. 1981. Kinetics of nitrate and ammonia uptake by the tropical fresh water macrophyte pista stratiotes L. Aquaculture 24, 11-19.

Negas, I., Alexis, M.N., Davies, S.J. and Petrichakis, G. 1995. Investigation to determine digestibility coefficients of various raw materials in diets for gilthead sea bream, *Sparus auratus* L. Aquacult. Res. 26, 185-194

Oron, G., Wildschut, L.R. and Porath, D. 1984. Wastewater recycling duckweed for protein production and effluent renovation. Water Sci. Technol. 17, 803-817.

Oron, G., Porath, D. and Wildschut, L.R. 1986. Wastewater treatment and renovation by different duckweed species. J. Envi. Eng. 112(2), 247-263.

Oron, G., Porath, D. and Jansen, H. 1987. Performance of duckweed species Lemna gibba on municipal wastewater effluent renovation and protein production. Biot. Bioeng. XXIX, 258-268.

Oron, G., De-Vegt, A. and Porath, D. 1988. Nitrogen removal and conversion by duckweed grown on wastewater. Water Res. 22(2), 179-184.

Oron, G. and Willers, H. 1989. Effect of wastes quality on treatment effecienct with duckweed. Water Sci. Technol. 21, 639-645.

Oron, G. 1990. Economic considerations in wastewater treatment with duckweed for effluent and nitrogen renovation. J. W. P. C. F. 62, 692-696.

Person-Le Ruyet, J., Delbard, C., Chartois, H. and Le Dellion, H. 1997. Toxicity of ammonia to turbot juveniles: 1-effects on survival, growth and food utilisation. Aquat. Living Resour. 10, 307-314

Pillay, T.V.R. 1992. Aquaculture and the environment, edited by Pillay T.V.R., first edition, 1992, pp. 189.

Pokorny, J. and Rejmankova, E. 1983. Oxygen regime in a fishpond with duckweed *(Lemnaceae)* and *Ceratophyllum*. Aquat. Bot. 17, 125-137.

Porath, D., Hepher, B. and Koton, A. 1979. Duckweed as an aquatic crop: evaluation of clones for aquaculture. Aquat. Bot. 7, 273-278.

Porath, D. and Pollock, J. 1982. Ammonia stripping by duckweed and its feasibility in circulating aquaculture. Aquat. Bot. 13, 125-131.

Prinsloo, J.F. and Schoonbee, H.J. 1987a. The use of sheep manure as nutrient with fish feed in pond fish polyculture in Transkei. Water SA 13(2), 119-123.

Prinsloo, J.F. and Schoonbee, H.J. 1987b. Investigations into the feasibility of a duck-fish-vegetable integrated agriculture-aquaculture system for developing areas in South Africa. Water SA, 13(2), 109-118.

Prinsloo, J.F. and Schoonbee, H.J. 1987c. Growth of the Chinese grass carp *Ctenopharyngodon idella* fed on cabbage wastes and kikuyu grass. Water SA, 13(2), 125-128.

Rangacharyulu, P.V., Mahendrakar, S.S., Mohanty, S.N., and Mukhopadhyay, P.K. 2003. Utilization of fermented silkworm pupae silage in feed for carps. Bioresource Technol. 86, 29-32

Rangeby, M., Johansson, P. and Pernrup, M. 1996. Removal of faecal coliforms in a wastewater stabilization pond system in Mindelo, Cape Verde. Water Sci.Technol. 34(11), 149-157

Reigh, R.C., Braden, S.L. and Craig, R.G. 1990. Apparent digestibility coefficients for common feed stuffs in formulated diets for red swamp Cray fish, *Procambarus clarkii*, Aquaculture 84, 321-334

Rejmankova, E. 1975. Comparison of Lemna gibba and *Lemna minor* from the production ecological viewpoint. Aquat. Bot. 1, 423-427.

Rifai, S. A. 1980. Control of reproduction of *Tilapia nilotica* using cage culture. Aquaculture 20, 177-185.

Robinson, E.H. and Li, M.H. 1994. Use of plant proteins in the catfish feeds: replacement of soybean meal with cottonseed meal and replacement of fishmeal with soybean meal and cottonseed meal. Journal of the World Aquaculture Society 25(2), 271-276

Rodrigues, A.M. and Oliveira, J.F.S. 1987. High-rate algal ponds: Treatment of wastewater and protein production:IV-Chemical composition of biomass produced from swine wastes. Water Sci. Technol. 19 (12), 243-248.

Rosegrant, M.W., Cali, X. and Cline, S.A. 2000. Global water outlook to 2025, averting and impending crisis. International Food Policy Research Institute, Washington, D.C., USA. September 2002.

Rusoff, L.L., Blakeney, E.W. Jr. and Culley, D.D. Jr. 1980. Duckweed (*Lemnaceae Family*): A potential source of protein and amino acids. J. Agric. Food Chem. 28, 848-850.

Sales, J. and Britz, P.J. 2001. Evaluation of different markers to determine apparent nutrient digestibility coefficient of feed ingredients for South African abalone (*Haliotis midae* L.). Aquaculture 202, 113-129

Sandbank, E. and Shelef, G. 1987. Harvesting of algae from high-rate ponds by flocculation-flotation. Water Sci. Technol. 19 (12), 257-263.

Sandbank, E. and van Vuuren, L.J. 1987. Microalgal harvesting by in situ autoflotation. Water Sci. Technol. 19 (12), 385-387.

Santos, E.J., Silva, E.H.B.C., Fiuza, J.M. Batista, T.R.O. and Leal, P.P. 1987. A high organic load stabilization pond using water hyacinth-a "BAHIA" experience. Water Sci. Technol. 19(10), 25-28.

Schillinkhout, A. and Collazos, C.J. 1992. Full-scale application of the UASB technology for sewage treatment. Water Sci. Technol. 25(7), 159-166.

Seghezzo, L., Zeeman, G., van Lier, J.B. Hamelers, H.V.M. and Lettinga, G. 1998. A review: the anaerobic treatment of sewage in UASB and EGSB reactors. Bioresour. Technol. 65, 175-190.

Schmitt, A.S.C. and Uglow, R.F.C. 1998. Metabolic responses of *Nephrops norvegicus* to progressive hypoxia. Aquat. Living Resour. 11(12), 87-92.

Seawright, D.E., Stickney, R.R. and Walker, R.B. 1998. Nutrient dynamics in integrated aquaculture-hydroponics systems. Aquaculture 160, 215-237.

Shehadeh, Z.H. and Feidi, I. 1996. Aquaculture development and resource limitation in Egypt. Aquaculture Newsletter No. 14.

Shelef, G., Azov, Y. and Moraine, R. 1982. Nutrients removal and recovery in a two-stage high-rate algal wastewater treatment system. Water Sci. Technol. 14, 87-100.

Shereif, M.M and Mancy, K.H. 1995. Organochlorine pesticides and heavy metals in fish reared in treated sewage effluents and fish grown in farms using polluted surface waters in Egypt. Water Sci. Technol. 32 (11), 153-161.

Shereif, M.M., Easa, M.E., El-Samra, M.I. and Mancy, K.H. 1995. A demonstration of wastewater treatment for reuse applications in fish production and irrigation in Suez, Egypt. Water Sci. Technol. 32 (11), 137-144.

Shireman, J.V., Colle, D.E. and Rottmann, R.W. 1977. Intensive culture of grass carp, Ctenopharyngodon idella in circular tanks J. Fish Biol. 11, 267-272

Singh, K., Garg, S.K., Kalla, A. and Bhatnagar, A. 2003. Oilcakes as protein sources in supplementary diets for the growth of Cirrhinus mrigala (Ham.) fingerlings: laboratory and field studies. Bioresource Technol. 86, 283-291

Sinha, V.R.P. 1999. Rural aquaculture in India "edited by Sinha V.R.P. and published by Regional Office for Asia and the Pacific (RAP), Food and agriculture Organization of the United Nations, Bangkok, Thailand, 1999.

Skillicorn, P., Spira, W. and Journey, W. 1993. Duckweed aquaculture " A new aquatic farming system for developing countries. The international bank for reconstruction and development / The World Bank, Washington DC., pp. 76.

Smith, M. D. and Moelyowati, I. 2001. Duckweed based wastewater treatment (DWWT): design guidelines for hot climates. Water Sci. Technol. 43(11), 291-299

Soliman, A.K., El-Horbeety, A.A.A., Essa, M.A.R., Kosba, M.A. and Kariony, I.A. 2000. Effects of introducing ducks into fishponds on water quality, natural productivity and fish production together with the economic evaluation of the integrated and non-integrated systems. Aquacult. Inter. 8, 315-326

Stott, R., Jenkins, T., Shabaan, M. and Mancy, E. 1997. A survey of the microbial quality of wastewater in Ismailia, Egypt and the implications for wastewater reuse. Water Sci. Technol. 35(11-12), 211-217.

Sullivan, J.A. and Reigh, R.C. 1995. Apparent digestibility of selected feedstuffs in diets for hybrid stripped bass (*Morone saxatilis* × *Morone chrysops*). Aquaculture 138, 313-322

Tacon, A.G.J. and Rodrigues, A.M.P. 1984. Comparison of chromic oxide, crude fiber, polyethyl and Acid-Insoluble Ash as dietary markers for the estimation of apparent digestibility coefficients in Rainbow trout. Aquaculture 43, 391-399

Thomas, S.L., and Piedrahita, R.H. 1997. Oxygen consumption rates of white sturgeon under commercial culture conditions. Aquacult. Eng. 16, 227-237.

Thomas, S.L. and Piedrahita, R.H. 1998. Apparent ammonia nitrogen production rates of white sturgeon (*Acipenser transmontanus*) in commercial aquaculture systems. Aquacult. Eng. 17, 45-55.

Tidwell, I.H., Webster, C.D., Sedlacek, J.D., Weston, P.A., Knight, W.L., Hill, Jr S.J., D'Abramo, L.R., Daniels, W.H., Fuller, M.J. and Montanez, J.L. 1995. Effects of complete and supplemental diets and organic pond fertilization on production of Macrobrachium rosenbergii and associated benthic macroinvertebrate populations. Aquaculture 138, 169-180

UNEP 2002. State of the environment and policy retrospective: 1972-2002. In Global Environment Outlook 3 Past, Present and Future Perspectives (Eds. Production team of UNEP), UNEP, London, UK.

Van Den Berg, G.W. and De La Noue, 2001. Apparent Digestibility comparison in rainbow trout (Oncorhyncus mykiss) assessed using three methods of faeces collection and three digestibility markers. Aquacult. Nutr. 7, 237-245

Van der Steen, P., Brenner, A. and Oron, G. 1998. An integrated duckweed and algae pond system for nitrogen removal and renovation Water Sci. Technol. 38(1), 335-343.

Van der Steen, P., Brenner, A., Shabtai, Y. and Oron, G. 2000. Effect of environmental conditions on faecal coliform decay in post-treatment of UASB-reactor effluent. Water Sci. Technol. 42(10-11), 111-118.

Van Dyke, J.M. and Sutton, D.L. 1977. Digestion of duckweed (*Lemna sp.*) by the grass carp (*Ctenopharyngodon idella*). J. Fish Biol. 11, 273-278

Vermaat, J.E. and Hanif, M.K. 1998. Performance of common duckweed species (Lemna gibba) and the water fern Azolla filiculoides on different types of wastewater. Water Res. 32(9), 2569-2576.

Vieira, S.M.M. and Garcia, Jr. A.D. 1992. Sewage treatment by UASB-reactor operation results and recommendations for design and utilization. Water Sci. Technol. 25 (7), 143-157.

Vieira, S.M.M., Carvalho, J.L., Barijan, F.P.O. and Rech, C.M. 1994. Application of the UASB technology for sewage treatment in a small community at Sumare, SaoPaulo state. Water Sci. Technol. 30(12), 203-210.

Virk, P. and Saxena, P.K. 2003. Potential of Amaranthus seeds in supplementary feed and its impact on growth in some carps. Bioresource Technol. 86, 25-27

Wallace, J.S. and Gregory, P.J. 2002. Water resources and their use in food production systems. Aquat. Sciences 64(4), 363-375

Wang, B. 1991. Ecological waste treatment and utilization systems on low-cost, energy-saving/ generating and resources recoverable technology for water pollution control in China. Water Sci. Technol. 24(5), 9-19.

Weatherup, R.N. and McCracken, K.J. 1998. Comparison of estimates of digestibility of two diets for rainbow trout, *Oncorhyncus mykiss* (Walbaum) using two markers and two methods of faeces collection. Aquacult. Res. 29, 527-533

WHO 1989. Health guidelines for the use of wastewater in agriculture and aquaculture. Technical report series, No. 778, World Health Organization, Geneva., pp. 77.

WHO and UNICEF 2000. Global Water Supply and Sanitation Assessment 2000 Report, pp 80

Williams, J., Bahgat, M., May, E., Ford, M. and Butler, J. 1995. Mineralisation and pathogen removal in gravel bed hydroponic constructed wetlands for wastewater treatment. Water Sci. Technol. 32 (3), 49-58.

Wong, M.H., Cheung, Y.H., Leung, S.F. and Wong, P.S. 1995. Reclamation of polluted river water for removal of nutrients by microalgae. Water Sci. Technol. 32 (3), 271-280.

Zaher, M., Begum, N.N., Hoo, M.E., Begum, M. and Bhuiyan, A.K.M.A. 1995. Suitability of duckweed, *Lemna minor* as an ingredient in the feed of tilapia, oreochromis niloticus. Bangladesh J. Zool. 23(1), 7-12.

Zhang, F.L., Zhu, Y. and Zhou, X.Y. 1987. Studies on the ecological effects of varying the size of fish ponds loaded with manure and feeds. Aquaculture 60, 107-116.

Zhao, J.H., Lam, T.T. and Guo, J.Y. 1997. Acute toxicity of ammonia to the early stage-larvae and juveniles of *Eriocheir sinensis* H.Milne-Edwards, 1853 (Decapoda: *Gaspidae*) reared in the laboratory. Aquacult. Res. 28, 517-525.

Zhiwen, S. 1999. The sustainable contribution of fisheries to food security in China. Sustainable contribution of fisheries to food security. Food and agriculture Organization of the United Nations, regional Office of Asia and the Pacific (FAO/RAP), RAP publication No. 1999/1-5. Edited by Leigh R., Kambhu and Chrlotte M. Menasveta.

Zhu, S., Chen, S., Hardy, R.W. and Barrows, F.T. 2001. Digestibility, growth and excretion response of rainbow trout (*Oncorhynchus mykiss* Walnaum) to feeds of different ingredient particle sizes. Aquacult. Res. 32, 885-893.

Zimmo, O.R., Al-Saed, R. and Gijzen, 2000. Comparison between algae-based and duckweed-based wastewater treatment: differences in environmental conditions and nitrogen transformations. Water Sci. Technol. 42(0-11), 215-222.

Zirschky, J. And Reed, .C. 988. The use of duckweed for wastewater treatment. J.W.P.C.F. 60, 1253-1255

Chapter Two

Nutrient recovery from domestic wastewater using a UASB-duckweed ponds system

Submitted to Bioresource Technology as:

Saber A. El-Shafai, Fatma A. El-Gohary, Fayza A. Nasr, N. Peter van der Steen and Huub J.Gijzen. Nutrient recovery from domestic wastewater using a UASB-duckweed ponds system.

Nutrient recovery from domestic wastewater using a UASB-duckweed ponds system

Abstract

The pilot-scale wastewater treatment system used in this study comprises a 40 l UASB reactor (6hours HRT) followed by three duckweed ponds in series (total HRT 15 days). During the warm season, the treatment system achieved removal values of 93%, 96% and 91% for COD, BOD and TSS, respectively. Residual values of ammonia, TKN and total phosphorus were 0.41 mgN/l, 4.4 mgN/l and 1.11 mgP/l, with removal efficiencies, 98%, 85% and 78%, respectively. The system achieved 99.998% faecal coliform removal during the warm season with final effluent containing 4×10^3 cfu/100ml. During the winter, the system was efficient in the removal of COD, BOD and TSS (92%, 93% and 91%) but not for nutrients (39% for ammonia, 53% for TKN and 57% for total P). The system is deficient in the removal of faecal coliform during the winter, producing effluent with 4.7×10^5 cfu/100ml. During the warm season, the N removal consisted of 80.5% by plant uptake, 4.7% by sedimentation and 14.8% by denitrification and volatilisation. Duckweed production rate of 33 ton dry matter per hectare per 8 months (warm season) was achieved. This amount of duckweed could replace wheat bran in commercial animal/fish feed with an equivalent value of 16,830 Egyptian Pound (US\$ 4,360, March 2002) based on local price of wheat bran.

Key words: wastewater, treatment, UASB, duckweed, production, faecal coliform, ammonia,

1. Introduction

Extension of affordable and manageable safe sanitation services for small towns and rural communities is needed. Therefore the objective of national and international policies is to accelerate expansion of environmentally friendly sanitation systems through the world, especially in the water scarce regions. The anaerobic treatment process is increasingly recognised as an important step of an advanced treatment technology for environmental protection and resource preservation and it represents, combined with other proper post-treatment methods, a sustainable and an appropriate wastewater treatment system for developing countries (Mergaert *et al.*, 1992; Seghezzo *et al.*, 1998). References of published work of, an Up-flow Anaerobic Sludge Blanket (UASB) reactor performance shows the advantages of this reactor. Application of the UASB reactor technology for domestic wastewater treatment is rapidly growing in developing countries (Draajer *et al.*, 1992; Schellinkhout and Collazos, 1992; Vieira and Garcia, 1992; Vieira *et al.*, 1994). The results of the UASB reactor showed the possibility of starting up the reactor for raw domestic sewage treatment without inoculation under temperature conditions ranging between 19 and 28 °C (Barbosa and Sant Anna, 1989). The UASB reactor, with a hydraulic retention time (HRT) of 7 hours, is an efficient primary treatment alternative to an anaerobic pond in Waste Stabilisation Pond (WSP) series (Dixo *et al.*, 1995). The UASB has a big potential in removing total suspended solids, biological oxygen demand and intestinal nematode eggs but is deficient in removing pathogenic bacteria and nutrients. The anaerobic treatment is a mineralization process, which provides nutrients-rich effluent, representing a good growth medium for the production of protein-rich aquatic plants.

The Nutrient rich effluent from the UASB should be post-treated, not to remove nutrients but to recover these nutrients. High rate algae ponds represent one of the most popular treatment methods and provide a huge amount of algae (Shelef *et al.*, 1982). The efficient recovery of the algae biomass however is complicated and only possible by using costly techniques like flocculation-flotation using chemicals (alum and ferric chloride) and centrifugation or combinations of these (Shelef *et al.*, 1982; Rodrigues and Oliveira, 1987; Sandbank and Shelef, 1987).

In aquatic macrophyte-based treatment systems, the sewage nutrients are recovered and transformed into easily harvested rich protein by-products. Recycling systems based on the treatment of municipal wastewater with protein production using duckweed represent a comprehensive solution (Cully and Epps, 1973; Oron *et al.*, 1988; Hammouda *et al.*, 1995). The duckweed has high productivity, high protein content, large nutrient uptake, easy handling, harvesting and processing (Oron *et al.*, 1984; Oron *et al.*, 1986; Hammouda *et al.*, 1995). Also, it has an extended growing period, low fibre content and reduces mosquito development (Shelef *et al.*, 1982; Abdalla *et al.*, 1987; Rodrigues and Oliveira, 1987; Santos *et al.*, 1987). The direct conversion of ammonia into plant protein in duckweed pond is a relatively energy efficient process compared to other alternative methods (Harvey and Fox, 1973; Oron *et al.*, 1987; Mbagwu and Adeniji, 1988; Zirschky and Reed, 1988). Under adequate operating conditions the duckweed pond provides both, a secondary effluent that meets the irrigation and aquaculture reuse criteria and an annual yield of about 55 tonne/ha, dry matter (Oron, 1990).

In recent years, research was focussed on duckweed and its role in wastewater treatment and its potential for nutrients recovery (Skillicorn *et al.*, 1993; Boniardi *et al.*, 1994; Ennabili *et al.*, 1998; Van der Steen *et al.*, 1999; Al-Nozaily *et al.*, 2000; Smith and Moelyowati, 2001; Cheng *et al.*, 2002). The process of treatment in duckweed covered sewage lagoons is a duckweed-mediated process, either directly through the nutrients recovery by plant uptake or indirectly by release of oxygen in the water column (Alaerts *et al.*, 1996). Microbial hydrolysis of the more complex organic N and P into NH_4^+ and ortho-PO_4^{3-} is the limiting step for enhancing biomass production. Providing adequate pre-treatment for sewage (like UASB) to release organically bound N and P might enhance the treatment efficiency (Alaerts *et al.*, 1996). The organic carbon represented in COD affects the efficiency of duckweed ponds (Oron *et al.*, 1987) and pre-treatment of wastewater in a settling cone for about 8 hours enhanced the ammonia uptake (Oron *et al.*, 1987). Thus the UASB may play a fundamental role in the improvement of duckweed pond performance.

Demonstration of a UASB-duckweed ponds system for domestic wastewater treatment and renovation in plant protein production is the aim of the present work. Investigation of the fate of influent nitrogen in the duckweed ponds represents the core of this study.

2. Materials and methods

An outdoor experiment was conducted at the department of Water Pollution Research and Control (National Research Centre), Cairo, Egypt. The experiment covered a period of one year, during which an intensive monitoring programme was carried out.

2.1. Up-flow Anaerobic Sludge Blanket (UASB) reactor (pre-treatment)

Domestic wastewater, collected from the neighbourhood, was pre-treated in a UASB reactor at 6 hours HRT. The reactor had a volume of 40 litre and was inoculated with anaerobic flocculent sludge at 25g total suspended solids (TSS) per litre. The reactor was fed continuously with raw sewage using a peristaltic dosing pump. Sludge removal was carried out regularly from the taps on the side walls of the reactor to keep the sludge height at two thirds of the reactor height

2.2. Duckweed ponds (post-treatment)

The duckweed species, *Lemna gibba* and *Lemna minor* are both naturally found in Egypt. The growth rate of Lemna *gibba* exceeds that of *Lemna minor* (Rejmankova, 1975; Porath *et al.*, 1979), while there is no difference in protein content and amino acids profile (Porath *et al.*, 1979). *Lemna gibba* was chosen as duckweed species in this study. A series of three duckweed ponds were coupled to the UASB reactor (Figure 1). Each pond occupied one square meter and was 0.48 m deep. The ponds were made of dark plastic sheet. Part of the UASB effluent (96 litre/day) was fed to the duckweed ponds, providing a total of 15 days hydraulic retention time (5 days in each pond). The duckweed ponds were inoculated with *Lemna gibba,* obtained from a local drain, at 600 grams fresh duckweed per m^2.

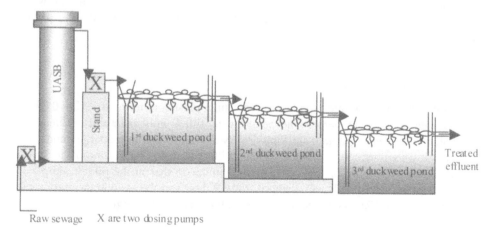

Fig. 1: Schematic diagram of the treatment system

2.3. Duckweed sampling

The plant growth rate and yield were monitored every 48 hours in each pond. The thickness of the residual cover after harvesting was maintained at 1000 g/m^2 (one layer). The harvested biomass was drained, weighed and dried in an oven at 70 °C. The dry matter content was calculated. The dry matter was powdered in a tissue grinder and 0.2 g was used for organic N analysis. The protein content was calculated based on: protein (g/g) = organic N (g/g) × 6.25 (Rusoff *et al.*, 1980). 0.1 g from the powder was taken and burned at 550 °C for one hour. The ash was analysed for phosphorus content using the per-sulphate digestion method (APHA, 1998) followed by the vanadomolybdate colorimetric method (APHA, 1998).

2.4. Sludge accumulation

Sludge accumulation measurements were carried out in all ponds in order to, better understand the fate of nitrogen in the system. The measurements were performed using plastic cups placed on the bottom in the centre of each pond. The surface area of each cup was about 71 cm^2. At the end of the measuring period (at day 186), the cups were removed from the ponds. Thickness of the sludge layer was measured and the sludge was analysed for dry solids concentration, and N and P content.

2.5. Ammonia volatilisation

Volatilisation of ammonia was measured by covering a small surface area of the ponds by using a glass bottle cut at its base. The mouth of the bottle was plugged with a piece of cork with a plastic tube passing through. The plastic tube was connected to a bottle containing boric acid. The bottles were immersed vertically in the ponds with the necks directed upwards. Collected gases were drawn through the boric acid solution, which was subsequently titrated with sulfuric acid to calculate the amount of ammonia present.

2.6. Evapotranspiration rate

The evapotranspiration rate was measured using plastic mini-ponds with 0.05 m^2 surface area and 30 cm depth. The ponds were filled with anaerobic effluent and inoculated with *Lemna gibba* at stocking density similar to that in the large ponds. The mini-ponds were operated as a batch system, with continuous harvesting of duckweed to keep the density more or less constant and similar to that in the pilot-scale ponds. The water in the mini-ponds was changed every 9 days and the evapotranspiration rate was measured as difference in the water level before and after changing.

2.7. Water sample analysis

Extensive monitoring of the treatment efficiency was performed by collecting weekly-samples from the different treatment units. All physico-chemical and microbiological analyses for pH, COD, BOD, ammonia N, nitrite N, nitrate N, Kjeldahl N, total P, TSS, hydrogen sulphide, dissolved oxygen and faecal coliform count were performed according to the Standard Methods (APHA, 1998).

3. Results

3.1. Characteristics of domestic wastewater

Analysis of the wastewater during the experimental period indicated that, COD, BOD and TSS were not significantly ($p > 0.05$) different in summer and winter period (Table 1). The ratio of COD:BOD was around 1:0.45 during the warm season and 1:0.42 during the winter. The average values for faecal coliform during summer and winter were 2.9×10^8 and 1.1×10^9 cfu/100ml, respectively. The TKN content of the raw wastewater was not significantly different ($p > 0.05$) between the two seasons. The only parameters, which were significantly different, are hydrogen sulphide ($p < 0.01$) and TP ($p < 0.05$) with higher values recorded in the warm season for H_2S and in the winter for TP.

3.2. Performance of the UASB

The efficiency of the UASB in the removal of COD, BOD and TSS was significantly ($p < 0.01$) reduced during the winter period. During the warm season, the UASB reactor achieved 79% removal efficiency for COD as well as 82% and 83% for the BOD and TSS, respectively. During the winter period, with temperature range of 13-18°C, these parameters were removed by 70%, 73% and 73%, respectively. The TKN and TP removal efficiencies represented 26% and 20% during the warm season and, 15 % and 28% during the wintertime. Faecal coliform removal is poor in the UASB reactor, with less than one log reduction for both seasons.

3.3. Removal efficiencies in duckweed ponds

Monitoring of the duckweed ponds during the year revealed that the total efficiency of the ponds in removing COD and BOD was not significantly ($p>0.05$) affected by the temperature. The TSS removal was significantly ($p < 0.01$) higher in winter than summer. The removal values for COD, BOD and TSS were, 64%, 73% and 43% in the warm season and, 72%, 75%, and 63% in winter.

Nutrient removal in the duckweed ponds was significantly ($p < 0.01$) reduced during the winter. The ammonia, TKN and TP removal efficiencies during the warm season were, 98%, 80% and 73%, respectively. These values decreased in winter to 44%, 45% and 40% for the ammonia, TKN and TP, respectively. The same pattern was observed for hydrogen sulphide concentrations. Hydrogen sulphide was negligible in the final effluent of the ponds during the warm season, despite the high concentration in the pond influent (3.6 mgS/l in winter and 8.4 mgS/l in summer). The faecal coliform count in the final pond effluent was 4×10^3 cfu/100ml during the warm season with an average removal of 99.93%. This count increased to 4.7×10^5 in winter, representing 99.7% removal.

Table (1): Characteristics of the raw sewage, UASB effluent and duckweed ponds effluent

Water sample	Raw wastewater		UASB effluent		Duckweed ponds effluent	
Parameter	Summer (June-October)	Winter (December-February)	Summer (June-October)	Winter (December-February)	Summer (June-October)	Winter (December-February)
Temperature (°C)	25-31	12.5-19	27-41	13-18	24-34	12.5 20
pH	6.6-7.5	6.8-7.2	6.8-7.4	6.9-7.1	7.2-8.3	7.5-7.9
COD (mgO$_2$/l)	749±193^{a*}	871±117^{a*}	151±44^{a**}	257±37^{b**}	49±18^{a**}	73±9^{b**}
BOD (mgO$_2$/l)	335±96^{a*}	362±35^{a*}	58±21^{a**}	99±13^{b**}	14±5^{a**}	25±6^{b**}
Ammonia N (mgN/l)	19.7±5.2^{a*}	17.2±2.2^{b*}	18.7±4.3^{a*}	18.8±2.3^{a*}	0.4±0.9^{a**}	10.4±1.5^{b**}
Nitrite N (mgN/l)	0.012±0.004	0.033±0.008	0.001±0.002	0.044±0.003	0.166±0.104	0.064±0.001
Nitrate N (mgN/l)	0.09±0.04	0.06±0.02	0.03±0.02	0.03±0.01	0.70±0.68	0.29±0.09
TKN (mgN/l)	29.6±5.8^{a*}	26.3±3.2^{a*}	21.8±4.3^{a*}	22.3±2.3^{a*}	4.4±1.4^{a**}	12.2±1.6^{b**}
Total P (mgP/l)	5.18±1.3^{a*}	6.16±1.04^{b*}	4.15±1.21^{a*}	4.42±0.77^{a*}	1.11±0.41^{a*}	2.69±0.76^{b*}
TSS (mg/l)	380±131^{a*}	330±54^{a*}	58±14^{a**}	83±12^{b**}	32±10^{a*}	31±5^{a*}
H$_2$S mgS/l	10.35±3^{a**}	2.9±0.85^{b**}	8.39±1.57^{a**}	3.57±0.75^{b**}	0	0.01±0.02
Faecal coliform (cfu/100ml)	2.9×10^8±2×10^8	1.1×10^9±8×10^8	8.9×10^7±6.5×10^7	3.2×10^8±3.2×10^8	4×10^3±3.7×10^3	4.7×10^5±5.5×10^5

* p > 0.05 in case of no significance difference or p < 0.05 in case of significance difference, ** p < 0.01 (significant difference between summer and winter values of parameter)

Table (2): Efficiency of the treatment system as % removal

Treatment Efficiency	UASB reactor		Duckweed ponds		Overall system	
Parameter	Summer	Winter	Summer	Winter	Summer	Winter
COD % removal	79 ± 5^a	70±1.8^b	64±17^a	72±1.3^a	93±4^a	92±0.4^a
BOD % removal	82±5^a	73±2^b	73±12^a	75±3^a	95±2^a	93±1^a
Ammonia N % removal	4±14	-9±16	98±4^a	44±7^b	98±3^a	39±10^b
TKN % removal	26±9	15±5	80±6^a	45±5^b	85±4^a	53±7^b
Total P % removal	20±9	28±5	73±8^a	40±8^b	78±7^a	57±7^b
TSS % removal	83±7^a	73±3^b	43±21^a	63±6^b	91±5^a	91±2^a
Faecal coliform % removal	63	73	99.93	99.7	99.998	99.94

p > 0.05, p < 0.01

41

3.4. Overall efficiency

Under Egyptian environmental conditions the system seems effective for the removal of COD, BOD and TSS during the whole year and for the recovery of nutrients and faecal coliform removal only during the warm season (about 8 months).

3.5. Duckweed biomass production and nutrient balances

Duckweed production rate is recorded in Table 3. During the warm season, duckweed harvesting was carried out every two days. Dry weight yield was about 138.8 Kg/ha/d in the first pond and, 135.8 Kg/ha/d and 126.4 Kg/ha/d in the two successive ponds. The average dry duckweed production during the warm season (8 months) is about 33 tons/ha. The estimated market price is around 16,830 LE (US$ 4,360), by comparing this biomass with the lowest price animal feed (wheat bran). The protein content of the dry matter achieved 22.3% for the first pond and decreased to 19.8% in the third pond. The TP in the dry matter of the duckweed ranged between 0.68% and 0.7% in the summer and between 0.84%-0.9% in the winter.

Table 3: Duckweed production rate in duckweed ponds and nutrients content as % of dry matter

Experimental period	Summer			Winter		
	1st dwp*	2nd dwp*	3rd dwp*	1st dwp*	2nd dwp*	3rd dwp*
Fresh yield (Ton/ha/d)	2.57±0.47	2.83±0.39	2.69±0.46	0.70±0.16	0.74±0.2	0.80±0.24
Dry yield (Kg/ha/d)	138.8	135.8	126.4	30.8	35.5	34.4
Dry matter (%)	5.4±0.2	4.8±0.4	4.7±0.5	4.4±0.5	4.8±0.2	4.3±0.4
Protein content	22.3±2.1	20.0±2.0	19.8±1.6	24.5±2.8	25.7±2.5	23.3±1.5
P content	0.70±0.04	0.69±0.03	0.68±0.09	0.89±0.04	0.90±0.02	0.84±0.03

* dwp is duckweed pond

Table 4: Sludge accumulation in the duckweed ponds during a period of 186 days in summer

Duckweed pond Parameter	1st duckweed pond	2nd duckweed pond	3rd duckweed pond
Total sludge weight (g)	245	286	327
Total sludge volume (l)	11.9	13.0	12.4
Height of sludge column (cm)	1.19	1.3	1.24
TKN in the sludge (g)	5.0 (2%)*	5.0 (1.8%)*	4.3 (1.3%)*
N sedimentation rate (mg N/d)	26.75	26.88	22.93

* Values between parentheses are % of TKN in the sludge as (TKN in grams/total sludge weight in grams) × 100

The sludge accumulation in the three duckweed ponds and its N content during the summer period is indicated in Table 4. The total nitrogen mass balance during the warm season is presented In Table 5. During the warm season, the total amount of nitrogen applied to the ponds was 2095 mgN/d (about 6.98KgN/ha/d). 78.5% of this load, i.e. 1645 mgN/d (5.48 KgN/ha/d) was removed in the duckweed ponds. The remaining amount was discharged in the effluent as organic nitrogen, nitrite and nitrate. In duckweed ponds, the amount of TN converted to plant protein represented about 81% of the total nitrogen removal, on average (4.42 KgN/ha/d). The TN lost via sedimentation represented about 4.7% of the total N removal. The unaccounted part represented 14.8% and is probably due to denitrification. Ammonia volatilisation was not detected by the applied procedure and is therefore assumed to be negligible. The total P recovered by duckweed ranged between 0.97 Kg P/ha/d in the first duckweed pond, 0.94 KgP/ha/d in the second pond and 0.86 KgP/ha/d in the third pond. In winter, the TN recovered by the plants was 1.21, 1.46 and 1.28 KgN/ha/d in the three duckweed ponds respectively. The TP recovery values were 0.27, 0.32 and 0.29 kgP/ha/d in the three duckweed ponds respectively.

Table 5: Daily fate of nitrogen in duckweed ponds during the summer

Duckweed pond	1st dwp	2nd dwp	3rd dwp	Overall
TN_{input} (mg N/day)	2095	1457	945	2095
$TN_{recovered}$ (mg N/day)	493 (77.3)[*]	435 (85)[*]	396.5 (80.1)[*]	1324.5 (80.5%)[*]
$TN_{sedimented}$ (mg N/day)	26.8 (4.2)[*]	26.9 (5.3)[*]	22.9 (4.6)[*]	76.5 (4.7%)[*]
$TN_{nitrified}$ (mg N/day)	21.1	76.3	-25.8	71.6
TN_{output} (mg N/day)	1457	945	450	450 (21.5%)[**]
$TN_{unaccounted}$ (mg N/day)	118.3 (18.5)[*]	50 (9.8)[*]	75.6 (15.3)[*]	243.9 (14.8%)[*]
$TN_{removed}$ (mg N/day)	638 (30.5%)[**]	512 (35%)[**]	495 (52.4%)[**]	1645 (78.5)[**]

[*] Values between parentheses are % from the removed nitrogen

[**] Values between parentheses are % from the influent nitrogen, - [$TN_{recovered}$ (total N recovered by duckweed) and $TN_{nitrified}$ (total N oxidised into nitrite and nitrate)]

4. Discussion

4.1. Raw sewage and UASB performance

Monitoring of the domestic sewage during one year indicated absence of high fluctuations in raw sewage composition between summer and winter. The ratio of COD:BOD was around 1:0.45 during the warm season and 1:0.42 in winter. These values are considered higher than normal value (1:0.67), (Elgohary et al., 1995). Faecal coliform concentrations recorded an average value of 2.9×10^8 cfu/100ml and 1.1×10^9 cfu/100ml in the warm and winter season, respectively. These values are in agreement with results obtained by Shereif et al. (1995) but higher than the results reported by El-Hamouri et al. (1995). The higher value in winter may be attributed to the low temperature and so low ageing and decay in the faecal coliform populations. It may also be attributed to the lower rate of water discharge in winter. Full-scale UASB reactors are being operated successfully in different cities (Draaijer et al., 1992; Schellinkhout and Collazes, 1992; Vieira and Garcia, 1992). Performance of the UASB reactor in this study is satisfactory, in accordance with the published data (Kirryama et al., 1992; Singh et al., 1996). In warm season, only limited reduction in the ammonia concentration was achieved, while the ammonia:TKN ratio increased from 67% in the raw sewage to 86% in the UASB effluent. In winter, the ratio increased from 65% to

84%. This is attributed to the conversion of organic nitrogen into ammonia under the anaerobic conditions, since about 70% of organic nitrogen turns into ammonia in the anaerobic reactors (Sanz and Polanco, 1990). The anaerobic process is a mineralisation process therefore only small amount of TKN and P was removed, 26% and 20% in warm season and, 15% and 28% in the winter for TKN and TP, respectively. White granules were detected in the UASB effluent, which are supposedly composed of elemental sulphur and attributed to the presence of sulphur reducing bacteria and sulphur re-oxidising bacteria. Sanz and Polanco (1990) concluded that, *Thiothrix* bacteria that were found in the anaerobic effluent are able to use hydrogen sulphide as energy source and transform it into sulphur, which is deposited as sulphur granules in their cell walls. The microbiological quality of the UASB effluent indicated that, 63% and 73% reduction in the faecal coliform had been achieved in the two successive periods. These are comparable with the results obtained by Lettinga (1992, 66%).

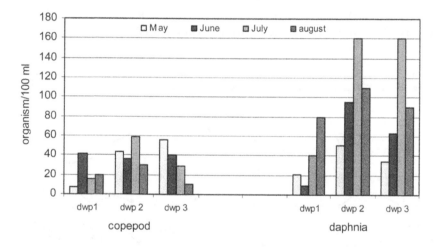

Fig. 2: Count of daphnia and copepod in duckweed ponds

4.2. Duckweed performance

Removal of COD, BOD and TSS from the anaerobic effluent fed to the three duckweed ponds reached about 64%, 73% and 43%, during the warm season and 72%, 75% and 63% in the winter, respectively. Oron *et al* (1987) observed COD and BOD removal values for *Lemna gibba* covered mini-ponds of about 63% and 92%, respectively at 20 days total HRT using settled sewage. Probably COD, BOD and TSS removal in duckweed ponds is mainly HRT dependent. At higher HRT a treatment system like duckweed ponds probably is not any more temperature dependent but it might be affected by temperature at low HRT. In this study the similarity in the pond efficiency in the removal of COD and TSS in summer and winter is mainly attributed to the presence of large number of small crustaceans, daphnia and copepods in the duckweed ponds in the warm season. Some of these crustaceans escape with the final effluent, Figure 2. These crustaceans are living organisms and therefore contribute to the COD and TSS but not to BOD test. In winter, the nitrification process was negligible in the duckweed ponds, Figure 3. In the warm season, the nitrification was limited in the first pond, probably because of the very low dissolved oxygen concentrations, Figure 4.

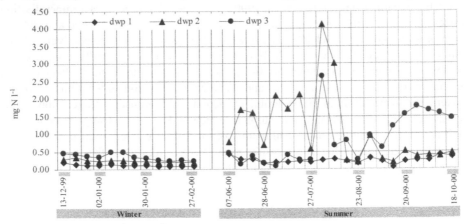

Fig. 3: Oxidised N (NO_2^{-1} + NO_3^{-1}) concentrations in duckweed ponds

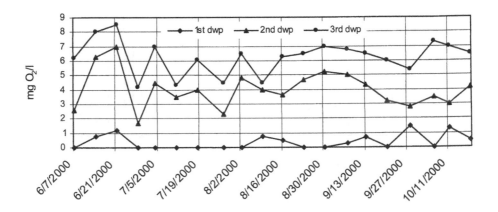

Fig. 4: Dissolved oxygen concentrations in duckweed ponds (summer)

4.3. Quality of duckweed biomass

In this study the dry matter content of the duckweed harvest ranged between 4.3% and 5.4%. This range is similar to what has been published in literature. The dry matter content of *Lemna* species ranged between 4.55 and 5.35% (Oron *et al.*, 1984; Oron *et al.*, 1987; Oron and Willers, 1989; Boniardi *et al.*, 1994; Ennabili *et al.*, 1998). The duckweed ponds achieved an average dry matter yield of 133.7 Kg/ha/d. By using *Lemna gibba* and domestic sewage, Oron and co-workers reported that the dry duckweed yield in sewage fed ponds ranged between 80 and 150 kg/ha/d (Oron *et al.*, 1984; Oron *et al.*, 1987; Oron and Willers, 1989). Van der Steen *et al* (1998) reported that the duckweed yield (*Lemna gibba*) was about 74-164 kg dry matter /ha/d. The protein content of the dry matter of duckweed represented 20.7% and 24.6% on average during the warm and cold season, respectively. During the monitoring of a full-scale duckweed-covered sewage lagoon, Alaerts *et al.* (1996) reported that the protein content of dry duckweed (*Spirodella*) ranged between 15.8% and 28%. Ennabili *et al.* (1998) reported a protein content of 17.8% in the dry matter of *Lemna gibba* grown on sewage while, Culley and Epps (1973) reported a range of 14%-25.9% in the dry matter of samples (*Lemna spp.*) collected from animal manure and human sewage contaminated lagoons in Louisiana and south-central Arkansas. Oron and co-workers reported a protein content of 15%-48.1% in the dry matter of *Lemna gibba* grown on sewage (Oron *et al.*, 1984; Oron *et al.*, 1987; Oron and Willers, 1989). Hammouda *et al.* (1995) reported higher protein content for *Lemna gibba* (31.8%-47.1%) grown on a mixture of Nile water and human sewage. The TP content in the dry matter of *Lemna gibba* in this experiment was 0.69% and 0.88%, on average during the summer and winter, respectively. In samples of *Lemna spp.* collected from animal waste lagoons and sewage contaminated lagoons, the TP content ranged between 0.57% and 0.8% (Cully and Epps, 1973). Alaerts *et al.* (1996) reported that S*pirodella species* harvested from sewage treatment lagoon contained TP in the range of 0.48% and 0.86% of dry matter biomass. Ennabili *et al.* (1998) reported an average value of 0.74% TP content in the dry matter of *Lemna gibba*.

4.4. Nutrients removal in duckweed ponds

In this treatment study, we confirmed that during the summer, the main part of TN removal (78.5%) was recovered by the duckweed (about 80.5%), while denitrification and sedimentation parts represented 14.8% and 4.7%, respectively. By using batch wastewater treatment system, Vermaat and Hanif (1998) found that about 50% of the TN input was unaccounted for. This amount was attributed to the nitrification-denitrification processes. In their study 34% of the TN was taken up by duckweed. The authors reported that during the experiment the pH increased to 9.6. This means that high concentration of algae and periphyton may have been present, which was ignored by the authors. The high pH value also enhances the process of ammonia volatilisation and this means that the duckweed pond in this case functioned similar to an algal pond. In case of algal ponds, the major ammonia removal route is considered to be volatilisation (Blier *et al.,* 1995; Silva *et al.*, 1995). Ammonia volatilisation is mainly linked to pH and temperature (Gomez *et al.*, 1995). In an integrated duckweed-algae system, treating UASB effluent Van der Steen *et al.* (1998) reported that 18% of influent N was recovered by duckweed, 3% was nitrified, 8% settled into the sediment and 73% was volatilised. The high value of the volatilised N was mainly attributed to the role of algal ponds in the system and not to the duckweed ponds (7 duckweed ponds and 3 algae ponds). In a full-scale duckweed-covered sewage lagoon (Alaerts *et al.*, 1996), the results revealed that 44% from the TN input has been recovered by the duckweed. Other parts of TN removal were, 28% seepage, 10% sedimentation, 9% as outflow and 9% has been unaccounted for. Alaerts *et al.* (1996) concluded that the plant consumed about 0.26 gN/m2/d and 0.05 gP/m2/d. In our study we got 0.44 gN/m^2/d and 0.092 gP/m^2/d. Al-Nozaily *et al.* (2000) reported that at ammonia concentration of 25 mg/l, the maximum N uptake by the duckweed *Lemna gibba* is about 4.55 KgN/ha/d (0.46 gN/m^2/d). Korner and Vermaat (1998) suggested that duckweed was responsible for 75% of the removed N while the remaining 25% were removed by other mechanisms, like denitrification and sedimentation. They added, the maximum amount of N and P that could be recovered by duckweed is about 0.59

gN/m²/d and 0.074 gP/m²/d. Zimmo *et al.* (2000) reported that over a period of 15 days and at experimental temperature of 20 °C, the N removal via denitrification and ammonia volatilisation represented 40% in duckweed ponds with initial ammonia concentration 50 mgN/l. In his experiment Zimmo reported nitrate concentrations of 11.9 and 14.8 mgN/l in spite of the low DO concentrations in the ponds (0.5-3 mgO₂/l).

4.5. Faecal coliform removal

Die-off of pathogenic bacteria is considered to be a complex phenomenon in waste stabilisation ponds. While detention time and temperature are considered to be the most influential parameters other factors have been reported by many investigators. These include the presence of predators, high pH value and solar radiation (Fernandez *et al.*, 1992; El-Hamouri *et al.*, 1995; Falowfield *et al.*, 1996; Rangeby *et al.*, 1996). The high pH value resulting from algal photosynthetic activity plays an important role in promoting faecal coliform die-off in the pond system (El-Hamouri *et al.*, 1995). In our present study, the removal of faecal coliform in duckweed ponds is more efficient in the warm season than in the winter. It is believed that in the current experiments the duckweed controlled the count of faecal coliform in the ponds through two main processes. Firstly, the recovery of nutrients from the pond may cause a deficiency in these nutrients, which might affect the faecal coliform removal. Secondly, the adsorption of the faecal coliform to the duckweed followed by harvesting might play a role in faecal coliform removal. But probably more important is that the duckweed cover protects the faecal coliform from solar radiation. Islam *et al.* (1990) reported that *Lemna minor* might serve as an effective environmental reservoir for *Vibrio cholera*. Dewedar and Bahgat (1995), reported that no decline in the counts was recorded in case of faecal coliform present in a set of dialysis sacs suspended under duckweed mat, whereas faecal coliform in dialysis bags exposed to sun showed a decline Van der Steen *et al.* (2000) reported that the faecal coliform decay in the dark parts of stabilisation ponds under conditions of carbon and nutrients sufficiency was expected to be negligible. They concluded that light mediated decay of faecal coliform was affected by pH, DO and solar radiation intensity. In addition the authors suggested that, the light attenuation by algae reduces the faecal coliform decay rate.

The overall efficiency of the system is promising, especially during the warm season (8 months in Egypt). The results are more encouraging than the recent published work of integrated UASB followed by two stabilisation ponds (Gosh *et al.*, 1999). Those authors reported that the treated effluent contained residual COD, ammonia and faecal coliform of 121 mg O₂/l, 27 mg N/l and 2.5×10⁵, respectively.

5. Conclusions

-The system is simple in operation, maintenance and therefore suitable for rural communities.
-The efficiency of the UASB in removing COD, BOD and TSS was lower during the winter. As a result the organic surface load to the duckweed ponds in winter was higher. However, this did not affect the removal efficiency in the duckweed ponds. Therefore the temperature did not significantly affect the overall efficiency of the system.

-The nutrient recovery from the UASB effluent in duckweed ponds is duckweed growth rate dependent. The duckweed plants captured about 80.5% of the removed nitrogen, while 4.7 % accumulated in the sediment and 14.8% was denitrified.

-The growth rate of duckweed was significantly reduced by lower temperature (13-20 °C) and hence the nutrient recovery decreased in winter.

-The faecal coliform removal in duckweed ponds is affected by the decline in temperature. Probably removal mechanisms in duckweed ponds are affected by temperature, nutrients availability and duckweed harvesting rate.

Acknowledgements

The authors would like to thank the Dutch government for financial support for this research within the framework of the "Wasteval" project. The project is a co-operation between the Water Pollution Control Department (NRC, Egypt), Wageningen University and UNESCO-IHE Institute for Water Education, The Netherlands.

References

Abdalla, A.L., Ambrosano, E.J., Vitti, D.M.S.S. and Silva, F.J.C. 1987. Water hyacinth (Eichhornia *crassipes*) in ruminant nutrition. Water Sci. Technol. 19(10), 109-112.

Alaerts, G.J., Mahbubar, M.D.R. and Kelderman, P. 1996. Performance analysis of a full-scale duckweed-covered sewage lagoon. Water Res. 30(4), 843-852.

Al-Nozaily, F., Alaerts, G. and Veenstra, S. 2000. Performance of duckweed-covered sewage lagoon. 2. Nitrogen and phosphorus balance and plant productivity. Water Res. 34 (10), 2734-2741

American Public health Association, 1998. Standard Methods for the Examination of Water and Wastewater, 20th edition, Washington.

Barbosa, R. A. and Sant Anna, G. L. jr, 1989. Treatment of raw domestic sewage in an UASB reactor. Water Res. 23, 1483-1490

Blier, R., Laliberte, G. and de la Noue, J. 1995. Tertiary treatment of cheese factory anaerobic effluent with *Phormidium bohneri* and *Micractinium puspillum*. Bioresource Technol. 22, 151-155

Boniardi, N., Vatta, G., Rota R., Nano, G. and Carra, S. 1994. Removal of water pollutants by Lemna gibba. The chemical Engireering Journal 54, 41-48.

Cheng, J., Bergamann, B.A., Classen, J.J., Stomp, A.M. and Howard, J.W. 2002. Nutrient recovery from swine lagoon water by *Spirodela punctata*. Bioresource Technol 81, 81-85.

Culley, D.D. Jr. and Epps, E.A. 1973. Use of duckweed for waste treatment and animal feed. J. W. P. C. F. 45 (2), 337-347.

Dewedar, A. and Bahgat, M. 1995. Fate of faecal coliform bacteria in a wastewater retention reservoir containing *Lemna gibba* L. Water Res. 29 (11), 2598-2600.

Dixo, N.G.H., Gambrill, M.P. Catunda, P.F.C. and van Haandel, A.C. 1995. Removal of pathogenic organisms from the effluent of an up flow anaerobic digester using waste stabilisation ponds. Water Sci. Technol. 31 (12), 275-284.

Draaijer, H., Maas, J.A.W., Schaapman, J.E. and Khan, A. 1992. Performance of the 5MLD UASB reactor for sewage treatment at Kanpur, India. Water Sci. Technol. 25 (7), 123-133.

El-Gohary, F., El-Hawarry, S., Badr, S. and Rashed,Y., 1995. Wastewater treatment and reuse for fish aquaculture. Water Sci. Technol. 32 (11), 127-136.

El-Hamouri, B., Jellal, J., Outabiht, H., Nebri, B., Khallayoune, K., Benkerroum, A., Hajli, A. and Firadi, R. 1995. The performance of a high-rate algal pond in the Moroccan climate. Water Sci. Technol. 31(12), 67-74

Ennabili, A., Ater, M. and Radoux, M. 1998. Biomass production and NPK retention in macrophytes drom wetlands of the Tingitan peninsula. Aquat. Bot. 62, 45-56.

Fallowfield, H.J., Cromar, N.J. and Evison, L.M. 1996. Coliform die-off rate constants in a high-rate algal pond and the effect of operational and environmental variables. Water Sci. Technol. 34(11), 141-147

Ferguson, A.R. and Bollard, E.G. 1969. Nitrogen metabolism of Spirodela oligorrhiza I-Utilization of ammonia, nitrate and nitrite. Planta (Berl) 88, 344-352.

Fernandez, A., Tejedor, C. and Chodi, A. 1992. Effect of different factors on the die-off of faecal bacteria in a stabilization pond purification plant. Water Res. 26(8), 1093-1098

Gomez, E., Casellas, C., Picot, B. and Bontoux, J. 1995. Ammonia elimination processes in stabilisation and high-rate algal pond systems. Water Sci. Technol. 31 (12), 303-312.

Gosh, C., Frijns, I. and Lettinga, G. 1999. Performance of silver carp (Hypophthalmicthys molitrix) dominated integrated post treatment system for purification of municipal wastewater in a temperate climate. Bioresource Technol. 69, 255-262

Hammouda, O., Gaber, A. and Abdel-Hameed, M.S. 1995. Assessment of the effectiveness of treatment of wastewater-contaminated aquatic systems with Lemna gibba. Enz. and Micro. Technol. 17, 317-323.

Harvey, R.M. and Fox, J.L. 1973. Nutrient removal using Lemna minor. J. Water P. C. F. 45 (9), 1928-1938.

Islam, S.I., Drasar, B.S. and Bradley, N.J. 1990. survival of toxigenic vibrio cholerae 01 with a common duckweed, Lemna minor in artificial aquatic ecosystems. Transaction of the Royal Society of Tropical Medicine and Hygiene 84, 422-424.

Kiriyama, K., Tanaka, Y. and Mori, I. 1992. Field test of a composite methane gas production system incorporating a membrane module for municipal sewage. Water Sci. Technol.. 25(7), 135-141

Korner, S. and Vermaat, J.E. 1998. The relative importance of Lemna gibba L., bacteria and algae for nitrogen and phosphorous removal in duckweed-covered domestic wastewater. Water Res. 33 (12), 3651-3661.

Lettinga, D., de Man, A., van der Last, A.R.M. Wiegan, W., van Knippenberg, K., Frings, J. and van Bauren, J.C.L. 1992. Anaerobic treatment of domestic sewage and wastewater. "Proceeding of first Middle East conference on water supply and sanitation for rural areas". Pp. 164-170, Cairo, Egypt.

Mbagwu, I.G. and Adeniji, H.A. 1988. The nutritional content of duckweed (Lemna paucicostata Hegelm) in the Kainji Lake area, Nigeria. Aquat. Bot. 29, 357-366.

Mergaert, K., van der Haegen, B. and Verstraete, W. 1992. Applicability and trends of anaerobic pre-treatment of municipal wastewater. Water Res. 26 (8), 1025-1033.

Nelson, S.G., Smith, B.D. and Best, B.R. 1981. Kinetics of nitrate and ammonia uptake by the tropical fresh water macrophyte Pista stratiotes L. Aquaculture 24, 11-19.

Oron, G., Wildschut, L.R. and Porath, D. 1984. Wastewater recycling by duckweed for protein production and effluent renovation. Water Sci. Technol. 17, 803-817.

Oron, G., Porath, D. and Wildschut, L.R. 1986. Wastewater treatment and renovation by different duckweed species. J. Envi. Eng. 112 (2), 247-263.

Oron, G., Porath, D. and Jansen, H. 1987. Performance of duckweed species Lemna gibba on municipal wastewater effluent renovation and protein production. Biot. Bioeng. XXIX, 258-268.

Oron, G., De-Vegt, A. and Porath, D. 1988. Nitrogen removal and conversion by duckweed grown on wastewater. Water Res. 22 (2), 179-184.

Oron, G. and Willers, H. 1989. Effect of wastes quality on treatment efficient with duckweed. Water Sci. Technol. 21, 639-645.

Oron, G. 1990. Economic considerations in wastewater treatment with duckweed for effluent and nitrogen renovation. J. W. P. C. F. 62, 692-696.

Porath, D., Hepher, B. and Koton, A. 1979. Duckweed as an aquatic crop: evaluation of clones for aquaculture. Aquat. Bot. 7, 273-278.

Porath, D. and Pollock, J. 1982. Ammonia stripping by duckweed and its feasibility in circulating aquaculture. Aquat. Bot. 13, 125-131.

Rangeby, M., Johansson, P. and Pernrup, M. 1996. Removal of faecal coliform in a wastewater stabilization pond system in Mindelo, Cape Verde. Water Sci.Technol. 34(11), 149-157

Rejmankova, E. 1975. Comparison of Lemna gibba and Lemna minor from the production ecological viewpoint. Aquat. Bot. 1, 423-427.

Rodrigues, A.M. and Oliveira, J.F.S. 1987. High-rate algal ponds: Treatment of wastewater and protein production: IV-Chemical composition of biomass produced from swine wastes. Water Sci. Technol. 19 (12), 243-248.

Rusoff, L.L., Blakeney, E.W. Jr. and Culley, D.D. Jr. 1980. Duckweed (*Lemnaceae Family*): A potential source of protein and amino acids. J. Agric. Food Chem. 28, 848-850.

Sandbank, E. and Shelef, G., 1987. Harvesting of algae from high-rate ponds by flocculation-flotation. Water Sci. Technol. 19 (12), 257-263.

Santos, E.J., Silva, E.H.B.C., Fiuza, J.M. Batista, T.R.O. and Leal, P.P. 1987. A high organic load stabilisation pond using water hyacinth-a "BAHIA" experience. Water Sci. Technol. 19 (10), 25-28.

Sanz, I. and Polanco, F. 1990. Low temperature treatment of municipal sewage in anaerobic fluidized bed reactors. Water Res. 24(4), 463-469

Schillinkhout, A. and Collazos, C.J. 1992. Full-scale application of the UASB technology for sewage treatment. Water Sci. Technol. 25 (7), 159-166.

Seghezzo, L., Zeeman, G., van Lier, J.B. Hamelers, H.V.M. and Lettinga, G. 1998. A review: the anaerobic treatment of sewage in UASB and EGSB reactors. Bioresource Technol. 65, 175-190.

Shelef, G., Azov, Y. and Moraine, R. 1982. Nutrients removal and recovery in a two-stage high-rate algal wastewater treatment system. Water Sci. Technol. 14, 87-100.

Shereif, M.M., Easa, M.E., El-Samra, M.I. and Mancy, K.H. 1995. A demonstration of wastewater treatment for reuse applications in fish production and irrigation in Suez, Egypt. Water Sci. Technol., 32 (11), pp. 137-144

Silva, S.A., de Oliveira, R., Soares, J., Mara, D.D. and Pearson, H.W. 1995. Nitrogen removal in pond systems with different configurations and geometries. Water Sci. Technol. 31(12), 321-330

Singh, K.S., Harada, H. and Viraraghavan, T. 1996. Low-strength wastewater treatment by an UASB-reactor. Bioresource Technol. 55,187-194

Skillicorn, P., Spira, W. and Journey, W. 1993. Duckweed aquaculture " A new aquatic farming system for developing countries. The International Bank for Reconstruction and Development / The World Bank, Washington DC.

Smith, M.D. and Moelyowati, I. 200 Duckweed based wastewater treatment (DWWT): design guidelines for hot climates. Water Sci. Technol. 43 (11), 291-299

Van der Steen, P., Brenner, A. and Oron, G. 1998. An integrated duckweed and algae pond system for nitrogen removal and renovation. Water Sci. Technol. 38 (1), 335-343.

Van der Steen, P., Brenner, A., Van Buuren, J. and Oron, G. 1999. Post-treatment of UASB reactor effluent in an integrated duckweed and stabilization pond system. Water Res. 33 (3), 615-620.

Van der Steen, P., Brenner, A., Shabtai, Y. and Oron, G. 2000. Effect of environmental conditions on faecal coliform decay in post-treatment of UASB-reactor effluent. Water Sci. Technol. 42 (10-11), 111-118.

Vermaat, J.E. and Hanif, M.K. 1998. Performance of common duckweed species (*Lemna gibba*) and the water fern Azolla filiculoides on different types of wastewater. Water Res. 32 (9), 2569-2576.

Vieira, S.M.M. and Garcia, Jr. A.D. 1992. Sewage treatment by UASB-reactor operation results and recommendations for design and utilisation. Water Sci. Technol. 25 (7), 143-157.

Vieira, S.M.M., Carvalho, J.L., Barijan, F.P.O. and Rech, C.M. 1994. Application of the UASB technology for sewage treatment in a small community at Sumare, SaoPaulo state. Water Sci. Technol. 30 (12), 203-210.

Zimmo, O.R., Al-Saed, R. and Gijzen 2000. Comparison between algae-based and duckweed-based wastewater treatment: differences in environmental conditions and nitrogen transformations. Water Sci. Technol. 42 (10-11), 215-222.

Zirschky, J. and Reed, S.C. 1988. The use of duckweed for wastewater treatment. J. Water P. C. F. 60, 1253-1261

Chapter Three

Effect of feeding strategy on tilapia (*Oreochromis niloticus*) production and water quality in duckweed-based fish aquaculture

Submitted to Bioresource Technology as:

Saber A. El-Shafai, Fatma A. El-Gohary, N. Peter van der Steen and Huub J.Gijzen. Effect of feeding strategy on tilapia (*Oreochromis niloticus)* production and water quality in duckweed-based fish aquaculture

Effect of feeding strategy on tilapia (*Oreochromis niloticus*) production and water quality in duckweed-based fish aquaculture

Abstract

Fresh and oven-dried (105 ^{0}C) duckweed (*Lemna gibba*) grown on domestic sewage were evaluated as sole feed source for Nile tilapia (*Oreochromis niloticus*) at a temperature range of 16-25 °C. The investigated daily feeding rates were 25% (dry), 50% (dry), 10% (fresh), 25% (fresh) and 50% (fresh). The feeding rates were calculated based on the fresh weight of duckweed and live body weight of fish. The results showed significantly ($p < 0.05$) higher SGR for tilapia fed on fresh duckweed than for those of parallel treatments fed on dry duckweed. In case of fresh duckweed, the SGR of tilapia fed on 10% feeding rate had a significantly ($p < 0.05$) lower value than those fed on 25% and 50%. There was no significant ($p > 0.05$) difference between the SGR of tilapia fed on 25% and 50% for both fresh and dry duckweed. The nitrogen mass balance showed that the highest nitrogen retention was in the treatment with 25% fresh duckweed, resulting in 53% assimilation of duckweed and 28% nitrogen retention from the nitrogen input. Under the experimental conditions applied, 25% fresh duckweed was regarded as the best feeding rate and feed form for tilapia.

Key words: fresh duckweed, oven-dried duckweed, *Lemna gibba*, domestic sewage, *Oreochromis niloticus*

1. Introduction

Cultured fish as an animal protein source for human consumption is considered a cheap animal protein source. Therefore aquaculture appears a promising protein source, although marine and fresh aquaculture only contribute about 12% of the current overall fish production. The production from aquaculture doubled between 1984 and 1991 while the increase of sea captures was 9% only (Rerat and Kaushik, 1995). In African countries, fish aquaculture production amounts to 0.03 Kg per capita per year, whereas fish intake is 10.5 Kg per capita per year, so the local or national production contributes 0.3% of the national demand (Rerat and Kaushik, 1995). In 1985, the total finfish aquaculture production in Africa was 11,550 tons annually, representing only 0.3% of global production (Huisman, 1985).

History of aquaculture is relatively recent in Africa compared to Asia; it is not familiar to the majority of the African countries. Most existing aquaculture systems have been introduced in Africa over the last 35 years. After classical fish farming technology was introduced from Europe in various African countries this development started to spread. The main farmed species is tilapia; in particular the farming of *Oreochromis niloticus* using organic-fertilised ponds or agriculture by-products is widespread. Use of mineral and feed pellets is very limited (Coche *et al.*, 1994).

There is a need to intensify the culture of tilapia all over Africa. One of the major constraints of fish farming development is the cost of fish feed which is an important part of the total costs involved in fish farming. This makes it essential to develop cheap and suitable diets either as feed supplement for ponds or as a complete feed for tanks and raceways. The cost of conventional fish feed ingredients, like fishmeal, soybean, wheat gluten and corn are high. This makes it necessary to investigate cheap and non-conventional protein sources for developing low-cost fish feed, suitable for use by small-scale farmers in rural fish farming areas. Because of high cost of fishmeal protein alternative ingredients from plant sources, agricultural wastes and by-products have been investigated (Appler, 1985; Robinson and Li, 1994; Davis *et al.*, 1995; Belal and Al-Dosari, 1999; Fontainhas *et al.*, 1999; Allan *et al.*, 2000; Davis and Arnold, 2000; Mbahinzireki *et al.*, 2001). The replacement of high cost plant protein ingredients by low cost ingredients was investigated as well (Robinson and Li, 1994; Belal, 1999; Falaye and Jauncey, 1999; Falaye *et al.*, 1999; Lim *et al.*, 2001). In recent years considerable attention was paid to some aquatic plants and wild grasses as alternative protein sources in the fish feed industry (Keke *et al.*, 1994; Zaher *et al.*, 1995; Basudha and Vishwanath, 1997; Fasakin and Balogun, 1998). Evaluation of the nutritional efficiency of diets-based on *Azolla* as substitute of fishmeal and other feed ingredients has been performed (Basudha and Vishwanath, 1997; Fasakin and Balogun, 1998). It was reported that the incorporation of duckweed in tilapia diet has an almost similar effect as rice bran and therefore, may be successfully used in intensive culture of tilapia (Zaher *et al.*, 1995). Considering the economic viability of the formulated feed, partial or complete replacement of the fishmeal with duckweed meal has been suggested. Varieties of aquatic plants are used as solely fish feed in experimental tanks and ponds stocked with herbivorous and/or omnivorous fish species (De Silva and Perera, 1983; Gaigher *et al.*, 1984; El-Sayed, 1992; Hassan and Edwards, 1992; Azim and Wahab, 1998; Azim *et al.*, 1998; Wahab *et al.*, 2001). Additional work still needed in this area of research by the use of aquatic plants grown in a nutrient rich environment like in pre-treated sewage. This study deals with the suitability of duckweed (*Lemna gibba*) grown on anaerobic pre-treated sewage as sole fish feed in static water systems and the effect of feeding rate on the growth of Nile tilapia. Comparison between fresh duckweed and oven-dried duckweed was studied.

2. Materials and methods

2.1. Fish species

Two different sizes of juvenile tilapia (*Oreochromis niloticus*) weighing 20 and 8 grams average weights were tested. The fish was brought from a local fish farm 25 km north of Cairo. The juveniles were adapted for two weeks prior to the experiment in three holding tanks. Each tank with an effective water volume of about 480 liter and 48 cm water depth. The small-size class was stocked in one tank at 100 fish stocking density while the large class-size was divided in two tanks at 50 fish stocking density in each tank. The tanks were fed with de-chlorinated tap water and aerated continuously using a compressor. During the adaptation period the fish was fed on a mixture of dry and fresh duckweed harvested from a UASB-duckweed pond system (El Shafai, submitted), treating domestic sewage.

2.2. Experimental design

Two sets of fish tanks, each five in number, were used in this experiment. The tanks had 0.25 m^2 surface area and 30 litre working capacity. One set of tanks was stocked with the small-class size at stocking density of five in each tank. The other set was stocked with large-class size at stocking density of 5 in each tank. The two sets were arranged into five parallel treatments. The first and second tanks were fed with dry duckweed at 25% and 50% daily feeding rate based on the fresh weight of the duckweed and live body weight of the fish. Duckweed was dried in an oven at 105 °C. The 3rd, 4th and 5th tank were fed on fresh duckweed at daily feeding rate of 10%, 25% and 50%, respectively (Table 1). Every week 15 litres from the tank water was removed and renewed by de-chlorinated tap water. The water in the treatment tanks was artificially aerated using a compressor.

Table 1: Operating conditions in fish tanks group A (large-size batch) and group B (small-size batch).

Item	Treatment 1	Treatment 2	Treatment 3	Treatment 4	Treatment 5
Surface area (m^2)	0.25	0.25	0.25	0.25	0.25
Working volume (l)	30	30	30	30	30
Initial mean body weight (g/fish), group A	19.94±2.17	22.84±2.83	20.6±1.52	17.58±1.43	17.08±0.9
Initial mean body weight (g/fish), group B	6.46±0.73	6.94±0.38	9.38±1.52	6.3±0.57	9.66±0.59
Stocking density (fish/tank)	5	5	5	5	5
Feeding rate *	25	50	10	25	50
Feed form	Dry	Dry	Fresh	Fresh	Fresh
Water changed (l/week)	15	15	15	15	15

* g fresh duckweed per 100 gram of live fish weight

2.3. Studied parameters and analytical procedures

The experiment was conducted for 90 days during which physical and chemical analysis of water in the treatment tanks was performed. The measurements covered temperature, pH, dissolved oxygen (DO), chemical oxygen demand (COD), total suspended solids (TSS), nitrite nitrogen, nitrate nitrogen, ammonia nitrogen, TKN (Total Kjeldahl Nitrogen) and total phosphorus. Water samples were collected weekly from the tanks just before the water refreshment. The pH, temperature and DO were measured daily. All the analyses were carried out according to the Standard Methods for the Examination of Water and Wastewater (APHA, 1998). The fish in each treatment were weighed monthly and the specific growth rates (SGR) were calculated based on the following equation:

$$SGR = \frac{(\ln Wf - \ln Wi)}{days} \times 100 \qquad (1)$$

Where Wi and Wf are initial and final mean body weight, respectively and days are the growth period

Total feed intake was recorded in each treatment and the feed conversion ratio (FCR) and protein efficiency ratio (PER) were calculated at the end of the trial based on the following equations:

$$FCR = \frac{\text{total feed ingested (dry weight)}}{\text{fish weight gain (wet weight)}} \qquad (2)$$

$$\text{Protein efficiency ratio (PER)} = \frac{\text{fish weight gain}}{\text{total protein ingested}} \qquad (3)$$

Proximal analysis of the duckweed composition was performed. Dry matter was calculated after drying in an oven at 105°C. The N and P content of the dry matter of the duckweed were measured. Proximal analysis of the fish material at the start and at the end of the experiment was performed. By the end of the trial, the accumulated sediment in each tank was measured and analysed for N content and N mass balance was made. The N content of duckweed, fish material and sediment was measured using the Kjeldahl method after acid digestion. The P content was measured colorimetricaly using the vanado-molybdate method after persulfate digestion.

2.4. Statistical Analysis

All the data are presented as mean values with standard deviations. Within the same group the SGRs of different treatments were subjected to one-way analysis of variance (one-way ANOVA). The parallel treatments between the two groups were also statistically analysed. The probability level of 95% was chosen to indicate significant and non-significant differences between the treatments.

3. Results

3.1. Water quality

The experiment was conducted over 90 days, during which the temperature range was 16-25 °C in all treatment tanks. The pH values and dissolved oxygen concentrations were within the desirable limit for Nile tilapia i.e. 7.2-8.4 for pH and above 5 mg O_2 l^{-1} for DO. Toxic nitrogenous compounds (ammonia and nitrite) did not accumulate in the tanks to toxic levels (Table 2). The results showed significantly (p<0.05) higher values of COD and TSS in the treatments of fresh duckweed compared to the parallel treatments of dry duckweed. The values also significantly increased with the increment of feeding rate, both for fresh and dry duckweed.

Table 2a: Water quality parameters in fish tanks, group A (large-size batch)

Parameter	Treatment 1	Treatment 2	Treatment 3	Treatment 4	Treatment 5
Temperature (°C)	16-25	16-25	16-25	16-25	16-25
pH	7.7-8.1	7.4-8	7.5-8.3	7.7-8.2	7.6-8.1
DO (mg O_2/l)	7.2±0.8	6.6±0.7	7.8±0.5	7.4±0.7	6.8±0.8
COD (mg O_2/l) [**]	63±6[b]	86±13[c]	43±13[a]	84±16[c]	109±17[d]
TSS (mg/l) [**]	39±6[a]	56±11[b]	30±9[a]	57±13[b]	76±13[c]
TAN[*] (mg N/l)	1.07±0.35	1.81±0.48	0.23±0.27	0.93±0.6	1.62±1.84
Nitrite (mg N/l)	0.14±0.04	0.16±0.03	0.09±0.04	0.14±0.05	0.15±0.03
Nitrate (mg N/l)	4.4±2.03	4.23±1.61	1.59±0.62	5.39±1.91	5.66±2.21
TKN (mg N/l)	8.2±1.6	12.2±2.4	4±0.8	8.5±3	14.6±2.8
TP (mg P/l)	1.56±0.52	2.06±0.52	0.9±0.25	1.89±0.31	2.16±0.19

[*]Total ammonia nitrogen

[**]Statistical comparison between the values in the same row and values with the same letter are not significant different (a, b, c and d)

Table 2b: Water quality parameters in fish tanks; group B (small-size batch)

Parameter	Treatment 1	Treatment 2	Treatment 3	Treatment 4	Treatment 5
Temperature (°C)	16-25	16-25	16-25	16-25	16-25
pH	7.6-8.3	7.2-8.2	7.5-8.4	7.5-8.2	7.7-8.1
DO (mg O_2/l)	7.4±0.5	7.5±0.6	8.2±0.4	8.2±0.5	7.7±0.7
COD (mg O_2/l) [**]	39±8[a]	58±10[b]	32±6[a]	73±13[c]	80±12[c]
TSS (mg/l) [**]	25±6[a]	35±4[b]	19±7[a]	41±12[c]	52±11[c]
TAN[*] (mg N/l)	0.6±0.31	1.26±0.36	0.05±0.14	0.22±0.31	0.39±0.48
Nitrite (mg N/l)	0.1±0.04	0.13±0.06	0.07±0.05	0.11±0.05	0.12±0.04
Nitrate (mgN /l)	1.52±0.76	1.57±0.56	0.91±0.28	1.57±0.65	3.79±1.8
TKN (mg N/l)	2.6±0.7	4±0.6	1.5±0.3	3.4±0.6	7.8±1.2
TP (mg P/l)	0.58±0.18	0.74±0.14	0.42±0.03	0.69±0.13	1.02±0.16

[*]Total ammonia nitrogen

[**]Statistical comparison between the values in the same row and values with the same letter are not significant different (a, b, c and d)

3.2. Growth performance

During the experimental period, the tanks receiving 50% fresh duckweed contained considerable amounts of duckweed floating on the surface in the morning of the day after feeding. This duckweed was collected, grinded and returned to the tanks. Statistical analysis of the SGRs obtained for the two groups showed no significant ($p > 0.05$) differences between the same treatments. The results of the SGRs of the different treatments were subjected to one way-ANOVA. The results showed significantly higher ($p < 0.05$) SGR in tilapia fed on 25% and 50% fresh duckweed than those of the parallel treatments fed on dry duckweed, (Table 3). In case of dry duckweed feed no significant ($p > 0.05$) difference was detected between the SGR of tilapia fed on 25% and those fed on 50% feeding rate. In case of fresh duckweed treatments, the SGR of tilapia fed on 10% feeding rate was significantly ($p < 0.05$) lower than the SGR of those fed on 25% and 50% feeding rate. No significant ($p > 0.05$) difference was present between the SGR of tilapia fed on 25% and those fed on 50% feeding rate.

Table 3a: Growth performance of tilapia in fish tanks, group A (large-size batch)

Item	Treatment 1	Treatment 2	Treatment 3	Treatment 4	Treatment 5
Initial mean body weight (g/fish)	19.94±2.17	22.84±2.83	20.6±1.52	17.58±1.43	17.08±0.9
Final mean body weight (g/fish)	24.68±2.7	28.26±2.67	25.7±1.87	28.76±1.4	27±2.02
Feed input (g dry feed/tank)	108.5	250	44.5	105.8	203.3
FCR (g dry feed/g fish)	4.6	9.2	1.7	1.9	4.1
SGR[*]	0.24±0.04[a]	0.24±0.04[a]	0.25±0.03[a]	0.55±0.04[b]	0.51±0.04[b]
PER	0.89	0.44	2.34	2.15	0.99
Solid waste (kg/ton fish)	910	1875	140	140	870

[*] Statistical comparison between the values in the same row and values with the same letter are not significant different (a and b)

Table 3b: Growth performance of tilapia in fish tanks, group B (small-size batch)

Item	Treatment 1	Treatment 2	Treatment 3	Treatment 4	Treatment 5
Initial mean body weight (g/fish)	6.46±0.73	6.94±0.38	9.38±1.52	6.3±0.57	9.66±0.59
Final mean body weight (g/fish)	8.06±0.92	8.68±0.57	11.98±1.36	10.3±0.71	15±0.89
Feed input (g dry feed/tank)	35.3	76	20.6	38.1	114.3
FCR (g dry feed/g fish)	4.4	8.7	1.6	1.9	4.2
SGR[*]	0.25±0.03[a]	0.25±0.03[a]	0.28±0.05[a]	0.55±0.04[b]	0.49±0.02[b]
PER	0.92	0.47	2.57	2.13	0.96
Solid waste (kg/ton fish)	890	1750	131	125	880

[*] Statistical comparison between the values in the same row and values with the same letter are not significant different (a and b)

3.3. Nitrogen mass balance and solid waste

By the end of the experiment the accumulated sediment in fish tanks was collected and analysed separately for dry matter and nitrogen content. The fish proximate analysis was performed for dry matter and nitrogen content as well. The total feed intake and its nitrogen content was calculated over the experimental period. Analysis of the quality of duckweed revealed a dry matter content of 4.5% and protein content of the dry matter of about 24.5%. All data were included in the nitrogen mass balance. The results presented in Table 4 showed that the percentage of the nitrogen content of the solid waste ranged between 4% and 10.5%. In case of optimal feeding rate (25% fresh duckweed) the value was 4.5% of dry matter. This amount of nitrogen represented 8% of the dietary nitrogen input. The nitrogen recovery by the fish was optimal in case of 25% fresh duckweed, representing 28-26% of the total nitrogen input. The major part of N loss was due to discharge with the effluent or due to storage in the sediment. An overview of the nitrogen contained in the waste products (sediment and effluent) is given in Table 5.

Table 4: Nitrogen mass balance in fish tanks

Item	Group A (large-size batch)					Group B (small-size batch)				
	T1	T2	T3	T4	T5	T1	T2	T3	T4	T5
N_{input}	4.26	9.8	1.74	4.15	7.97	1.39	2.97	0.81	1.5	4.48
(g N/ tank)										
$N_{disc.}$	2.26	2.95	0.974	2.46	3.71	0.76	1.02	0.45	0.9	2.18
(g N/ tank)	(53)	(30)	(56)	(59)	(46.5)	(54.5)	(34)	(56)	(60)	(49)
$N_{sed.}$	1.28	5.08	0.14	0.35	2.77	0.39	1.59	0.06	0.113	1.203
(g N/ tank)	(30)	(52)	(8)	(8)	(35)	(28)	(53)	(7)	(7.5)	(27)
$N_{reco.}$	0.488	0.559	0.526	1.15	1.02	0.155	0.168	0.256	0.387	0.517
(g N/ tank)	(11.5)	(6)	(30)	(28)	(13)	(11)	(6)	(32)	(26)	(11.5)
$N_{unacc.}$	0.232	1.211	0.1	0.19	0.47	0.085	0.192	0.044	0.09	0.58
(g N/ tank)	(5.5)	(12)	(6)	(5)	(6)	(6)	(6.5)	(5)	(6)	(13)

Values between brackets is % from N input

$N_{disc.}$ total nitrogen discharged in the effluent, $N_{sed.}$ nitrogen content in the sediment, $N_{reco.}$ nitrogen recovery in fish biomass, $N_{unacc.}$ unaccounted for nitrogen

Table 5: Nitrogen-containing waste material in fish tanks

Item	Group A					Group B				
	T1	T2	T3	T4	T5	T1	T2	T3	T4	T5
$N_{disc.}$ (kg/ton fish)	95.4	108.9	38.2	44	79.6	95	117.2	34.6	45	81.6
$N_{sed.}$ (kg/ton fish)	54.6	187.5	5.6	6.3	55.7	49	183.8	4.6	5.6	44
TN (g N/100 g dry sediment)	6	10	4	4.5	6.4	5.5	10.5	3.5	4.5	5
Total wasted-nitrogen (kg/ton fish)	150	296.4	43.8	50.3	135.3	144	301	39.2	50.6	125.6

$N_{disc.}$ total nitrogen discharged in the effluent and $N_{sed.}$ Total nitrogen in the sediment

4. Discussion

4.1. Water quality

Higher values of COD and TSS in the fresh duckweed compared to the dry duckweed feeding trials could be attributed to two factors. Firstly, debris of grinded duckweed that remains from the previous feeding. For the treatment with 50% feeding rate as fresh duckweed, the fresh duckweed remained till the next day on the water surface. This duckweed was collected, grinded and returned to the pond, in this way affecting the water quality. Secondly, it was also noticed that faeces debris in case of fresh duckweed tended to suspend in the water phase and big particles floated on the surface of the water. In case of dry duckweed the faeces settled to the bottom. In the trial with 50% feeding rate the dry duckweed floated on the pond surface and settled to the bottom after a while. Higher feeding rate (50%) in both fresh and dry duckweed negatively affected the water quality. Increasing the feeding rate resulted in high production of wastes either soluble or non-soluble.

4.2. Growth performance

The results of the growth performance of tilapia showed maximum SGR of 0.55 at 25% feeding rate on fresh duckweed. This result lies within the reported values of 0.51 to 0.73 for tilapia fed on formulated feed containing fishmeal, blood meal and water hyacinth leaf as substitute for groundnut cake at ratios of 0, 15, 30, 45 and 60% (Keke *et al.*, 1994). These results were obtained with fish of 5 gram initial body weight and temperature range of 21-25 °C. This might explain the higher maximum SGR (0.73) reported by the author in comparison to our findings with fish of 8 and 20 grams and at temperature range of 16-25°C. The higher maximum value could also be attributed to essential elements present in the fishmeal of the diet. It is known that small fish have higher growth rates than larger ones, although in our experiment no differences between the 20 and 8 gram tilapia were observed. This could be attributed to higher sensitivity of small fish to the low temperature range that has been used in our experiment. Zaher *et al.* (1995) got SGR of 1.8-2.2 for tilapia weighing 2.5 gram and fed on three fishmeal-based iso-nitrogenous diets containing water hyacinth, rice bran or duckweed. These higher values might be attributed to the small size of tilapia and some essential elements that are present in fishmeal. The results in the current study are better than the SGR (0.17) of catfish fed on fishmeal-based diet containing 100% *Azolla* as substitute of soybean protein (Fasakin and Balogun, 1998). Gaigher *et al.* (1984) recorded SGR of 0.63-0.72 by using hybrid tilapia weighing 2.7 grams and fed on fresh duckweed (*Lemna gibba*) at temperature range of 24-30 °C. Hassan and Edwards (1992) conducted a similar experiment in static water concrete tanks, using fresh *Lemna gibba* and *Spirodela sp.* at temperature range of 20-35°C and Nile tilapia weighing 25 grams. *Lemna gibba* provided better results than *Spirodela sp.* (SGR of 0.41, 0.8, 1.34, 1.4, 1.1 and 1.03 at feeding rates of 1, 2, 3, 4, 5 and 6% based on the dry matter content of duckweed). The current experiment seems to give slightly better results since the SGR was 0.55 at 25% feeding rate of fresh duckweed (1.13% based on dry matter). The optimal feeding rate of 25% was probably limited by the low temperature, which dropped to 16 °C. The results for *Lemna gibba* here are more promising than the published data of SGR for adult tilapia (40 grams) fed on formulated feed containing 100% *Azolla* as substitute for fishmeal (0.13) as well as values obtained (-0.11) in case of fresh *Azolla* alone (El-Sayed, 1992). The better results of *Lemna gibba* in comparison to *Azolla* most likely are due to the high fibre content and low protein content in *Azolla*.

The FCR was 1.9 at optimum (25% fresh duckweed) feeding rate while PER was 2.13. Basudha and Vishwanath (1997) reported FCR and PER of 1.75 and 3.8 for medium carp fed on fishmeal-based diet containing some *Azolla* powder whereas the values were 2 and 3.2 in case of feed containing rice bran and mustard oil cake. Zaher *et al.* (1995) obtained FCR values of 2.1-2.5 for tilapia fed on formulated feed containing water hyacinth, duckweed or rice bran. The results of FCR and PER in the current experiment showed potential advantage of *Lemna gibba* over other plants and agricultural wastes and by-products. Other authors investigated that the substitution of plant protein with more than 10% cocoa husk resulted in significantly lower growth performance and high FCRs, which ranged between 2.1-7.8 (Falaye *et al.*, 1999; Falaye and Jauncey, 1999). Cocoa husk has low protein content and very high fibre content, which negatively affects the growth rate and protein efficiency. The FCR obtained with duckweed feed is better than 3.6-3.9 that has been reported for the filamentous algae *Hydrodictyon reticulum* as fishmeal substitute (Appler, 1985). The FCR of fresh duckweed, *Lemna gibba,* was in the range of 1.9-3.3 for tilapia (Gaigher *et al.*, 1984; Hassan and Edwards, 1992). The reported value for Nile tilapia (13.5 g) fed on formulated feed containing duckweed *Spirodela sp.,* as fishmeal substitute was 1.6-4.3 (Fasakin *et al.*, 1999). Protein efficiency ratio of tilapia fed on fishmeal-based diet with barley seed as replacement of dietary corn was reported to be in the range of 2.5-2.9 (Belal, 1999), which is better than our results. This might be attributed to some essential growth factor elements in the fishmeal. In Denmark, the FCR in fresh water fish farms supplied with high quality commercial feed, ranged between 1.05-1.33 (Iversen, 1995).

The growth performance in this study showed significantly higher growth rates in fresh duckweed treatments than in the dry duckweed treatments. All experiments carried out by other authors on duckweed-fish aquaculture, used fresh duckweed as sole feed for tilapia (Gaigher *et al.*, 1984; Hassan and Edwards, 1992). No previous work compared between the fresh and dry duckweed. Opposite to our findings for duckweed, El-Sayed (1992) reported that dry water fern is better than fresh water fern in rearing tilapia. This result was attributed to the high moisture content of the fresh plant, causing limitation in the plant intake. Gaigher *et al.* (1984) reported that the intake of fresh duckweed was limited by the large bulk of duckweed caused by air pockets and moisture content. The high fibre content of the plant material was proposed as the main reason of low digestibility (Falaye *et al.*, 1999; Hassan and Edwards, 1992) while Fasakin and Balogun (1998) proposed the presence of some anti-nutritional factors in *Azolla*. The higher SGR obtained with fresh duckweed in the current experiment may have to do with the moisture content of the feed. The moisture content might have a role in enhancing the digestibility of the feed. Tacon *et al.* (1995) reported that moist pellet with 39% moisture content provided better SGR and FCR and less solid wastes in culture of sea bass (*Lates calcarifer*) and grouper (*Epinephelus tauvina*). In cage culture of tilapia there was significant higher SGR and net yield in case of tilapia fed on moist pellets than those fed on dry pellets (Guerrero, 1980), which support the hypothesis that presence of some water in the feed might enhance the hydrolysis process in the fish intestine. In addition, the drying process might cause the loss of some essential elements and this may also explain the lower SGR observed in the treatments with dry duckweed. Sewage bacteria in the fresh duckweed might enhance the microbial decomposition of duckweed in the fish intestine. Microbial decomposition of duckweed (*Lemna gibba*) was three folds higher in the presence of sewage bacteria and the time required to lose half of the organic matter of duckweed in the presence of sewage bacteria was one third of that in the absence of sewage bacteria (Szabó *et al.*, 2000).

4.3. Nitrogen mass balance

The nitrogen mass balance showed 28% net nitrogen recovery and 60% and 8% for nitrogen discharge and sediment accumulation at optimum feeding rate. Gaigher *et al.* (1984) reported about 49% net nitrogen recovery and 37% discharge in the effluent for experiments with duckweed as fish feed. The low value of net recovery in the current experiment and the high value of discharged nitrogen are mainly attributed to the low temperature, which decrease metabolic rate and raised metabolic energy requirements. N is the main energy source for fish and at low temperature fish catabolise more N to save its body temperature. The results published by Gomes *et al.* (1995) showed that about 4.3-13.7% from the total nitrogen-input as lose in the sediment (faeces). The same authors reported that in case of fishmeal-based diets and plant protein-based diets the net nitrogen recovery was in the range of 15.5-30% of the total dietary nitrogen. Nitrogen accumulation in the sediment of the tanks was higher in case of dry duckweed in comparison to the fresh duckweed, which is mostly attributed to the low digestibility of dry duckweed and sedimentation of considerable part of dry duckweed at the higher feeding rates. The total nitrogen waste in the optimum feeding rate in this trial was 50.3 kg N/ton fish production. The nitrogen accumulated in the sediment represented 6.3 and the nitrogen in the effluent represented 44 kg N/ton fish; these values are comparable to the results obtained by the above-mentioned authors. Tacon *et al.* (1995) reported that the production of one ton of carp release about 53-71 kg N while in the typical Nordic fish farm 55 kg N/ton is released per ton of fish production (Enell, 1995). By replacing the fishmeal with plant protein ingredients the results showed that the solid and soluble N was in the range of 6.3-16.1 and 55-113 kg N for each ton of fish production (Gomes *et al.*, 1995). The solid waste of 140 kg per ton fish is comparable to what has been mentioned (120-240 Kg) by Rodrigues (1995).

5. Conclusions

-It is possible to raise tilapia (*Oreochromis niloticus*) with duckweed (*Lemna gibba*) grown in pre-treated domestic sewage as the only feed source in pond or static water system.

-The use of fresh duckweed is superior to oven dried (105 °C) duckweed in rearing tilapia in static water bodies. This might be attributed to the sewage bacteria and/or moisture content of the fresh duckweed that may have increased the digestibility. Furthermore, reduced digestibility of dry duckweed due to drying process might be a reason.

-The optimum feeding rate at the temperature range of 16-25 °C is 25% based on the fresh weight of duckweed.

-The waste production per ton of fish production was lower in case fresh duckweed was used as feed, as compared to the situation with dry duckweed as feed.

Acknowledgements

The authors would like to thank the Dutch government for financial support for this research within the framework of the SAIL funded "Wasteval" project (LUW/MEA/971). The project is a co-operation between the Water Pollution Control

Department (NRC, Egypt), Wageningen University and UNESCO-IHE Institute for Water Education, The Netherlands. The authors would like to express their sincere thanks to the technicians of the Water Pollution Control Department, NRC, Egypt for analytical support.

References

American Public Health Association 1998. Standard Methods for the Examination of Water and Wastewater, 20th edition, Washington.

Allan, G.L., Rowland, S.J., Mifsud, C., Glendenning, D., Stone, D.A.J. and Ford, A. 2000. Replacement of fishmeal in diets of Australian Silver perch, *Bidyanus bidyanus* V. least-cost formulation of practical diets. Aquaculture 186, 327-340

Appler, H.N. 1985. Evaluation of *Hydrodictylon reticulatum* as protein source in feeds for *Oreochromis niloticus* and *Tilapia zillii*. J. Fish Biol. 27, 327-334

Azim, M.E. and Wahab, M.A. 1998. Effects of duckweed (*Lemna sp.*) on pond ecology and fish production in carp polyculture of Bangladesh. Bangladesh J. Fish. 21(1), 17-28

Azim, M.E., Wahab, M.A., Haque, M.M., Wahid, M.I. and Haq, M.S. 1998. Suitability of duckweed (*Lemna sp.*) as dietary supplement in four species polyculture. Progress. Agric. 9(1 & 2), 263-269

Basudha, C. and Vishwanath, W. 1997. Formulated feed based on aquatic weed azolla and fishmeal for rearing carp *Osteobrama belangeri* (Valenciennes). J. Aquacult. Trop. 12, 155-164.

Belal, I.E.H. 1999. Replacing dietary corn with barley seed in Nile tilapia, *Oreochromis niloticus* (L.), feed. Aquacult. Res. 30, 265-269

Belal, I.E.H. and Al-Dosari, M. 1999. Replacement of fishmeal with Salicornia meal in feeds for Nile tilapia (*Oreochromis niloticus*). J. World Aquacult. Soc. 30(2), 285-289

Coche, A.G., Haight, B.A. and Vincke, M.M.J. 1994. Aquaculture development and research in sub-Saharan Africa "CIFA technical paper, FAO, Rome.

Davis, D.A., Jirsa, D. and Arnold, C.R. 1995. Evaluation of soybean proteins as replacements for menhaden fish meal in practical diets for the red drum *Sciaenops ocellatus*. J. World Aquacult. Soc. 26(1), 48-57

Davis, D.A. and Arnold, C.R. 2000. Replacement of fishmeal in practical diets for the pacific white shrimp, *Litopenaeus vannamei*. Aquaculture 185, 291-298

De Silva, S.S. and Perera, M.K. 1983. Digestibility of an aquatic macrophyte by the *Cichlid etroplus* suratensis (Bloch) with observations on the relative merits of three indigenous components as markers and daily changes in protein digestibility. J. Fish Biol. 23, 675-684

El-Shafai, S.A., El-Gohary, F.A., Nasr, F.A., Peter van der Steen N. and Gijzen, H. J. (Submitted) Nutrient recovery from domestic wastewater using a UASB-duckweed ponds system.

El-Sayed, A.F.M. 1992. Effects of substituting fish meal with *Azolla pinnata* in practical diets for fingerling and adult Nile tilapia, *Oreochromis niloticus* L. Aquacult. Fish. Manage. 23, 167-173

Enell, M. 1995. Environmental impact of nutrients from Nordic fish farming. Water Sci. Technol. 31(10), 61-71

Falaye, A.E. and Jauncey, K.C. 1999. Acceptability and digestibility by tilapia Oreochromis niloticus of feed containing cocoa husk. Aquacult. Nutr. 5, 157-161

Falaye, A.E., Jauncey, K. and Tewe, O.O. 1999. The growth performance of tilapia (*Oreochromis niloticus*) fingerlings fed varying levels of cocoa husk diets. J. Aquacult. Trop. 14(1), 1-10

Fasakin, A.E. and Balogun, A.M. 1998. Evaluation of dried water fern (*Azolla pinnata*) as a replacer for soybean dietary components for *Clarias gariepinus* fingerlings. J. Aquacult. Trop. 13(1), 57-64

Fasakin, E.A., Balogun, A.M. and Fasuru, B.E. 1999. Use of duckweed, *Spirodela polyrrhiza* L. schleiden, as a protein feed stuff in practical diets for tilapia, *Oreochromis niloticus* L. Aquacult. Res. 30, 313-318

Fontainhas, F.A., Gomes, E., Reis-Henriques, M.A. and Coimbra, J. 1999. Replacement of fishmeal by plant proteins in the diet of Nile tilapia: digestibility and growth performance. Aquacult. Int. 7, 57-67

Gaigher, I.G., Porath, D. and Granoth, G. 1984. Evaluation of duckweed (*Lemna gibba*) as feed for tilapia (*Oreochromis niloticus* × *Oreochromis aureus*) in a recirculating unit. Aquaculture 41, 235-244.

Gomes, E.F., Rema, P., Gouveia, A. and Teles, A.O. 1995. Replacement of fishmeal by plant proteins in diets for Rainbow trout (*Oncorehynchus mykiss*): effect of quality of fishmeal based control diets on digestibility and nutrient balances. Water Sci. Technol. 31(10), 205-211

Guerrero, R.D.III. 1980. Studies on the feeding of *Tilapia nilotica* in floating cages. Aquaculture, 20, 169-175

Hassan, M.S. and Edwards, P. 1992. Evaluation of duckweed (*Lemna perpusilla* and *Spirodela polyrrhiza*) as feed for Nile tilapia (*Oreochromis niloticus*). Aquaculture 104, 315-326.

Huisman, E.A. 1985. Current status and role of aquaculture with special reference to the African region. Aquaculture Research in the African region "Proceeding of the African Seminar on aquaculture, organized by the IFS, 7-11 October 1985, pp. 11-22

Iversen, T.M. 1995. Fish farming in Denmark: Environmental impact of regulative legislation Water Sci. Technol. 31(10), 73-84

Keke, I.R., Ofajekwu, C.P, Ufodike, E.B. and Asala, G.N. 1994. The effect of partial substitution of groundnut cake by water hyacinth (*Eichhornia crassipes*) on growth and food utilisation in the Nile tilapia, *Oreochromis niloticus* (L.). Acta Hydrobiol. 36, (2), 235-244

Lim, H.A., Ng, W.K., Lim, S.L. and Ibrahim, C.O. 2001. Contamination of palm kernel meal with *Aspergillus flavus* affects its nutritive value in pelted feed for tilapia, *Oreochromis mossambicus*. Aquacult. Res. 32, 895-905

Mbahinzireki, G.B., Dabrowski, K., Lee, K.J., El-Saidy, D., and Wisner, E.R. 2001. Growth, feed utilisation and body composition of tilapia (*Oreochromis sp.*) fed with cottonseed meal-based diets in a recirculation system. Aquacult. Nutr. 7, 189-200

Rerat, A. and Kaushik, S.J. 1995. Nutrition, animal production and the environment. Water Sci. Technol. 31(10), 1-19

Robinson, E.H. and Li, M.H. 1994. Use of plant proteins in the catfish feeds: replacement of soybean meal with cottonseed meal and replacement of fishmeal with soybean meal and cottonseed meal. J. World Aquacult. Soc. 25(2), 271-276

Rodrigues, A.M.P. 1995. Biological and nutritional approach to the environmental impact of trout culture in Portugal. Water Sci. Technol. 31(10), 239-248.

Szabó, S., Braun, M., Nagy, P., Balazsy, S. and Reisinger, O. 2000. Decomposition of duckweed (*Lemna gibba*) under axenic and microbial conditions: flux of nutrients between litter water and sediment, the impact of leaching and microbial degradation. Hydrobiologia, 434, 201-210.

Tacon, A.G.J., Phillips, M.J. and Barg, U.C. 1995. Aquaculture feeds and the environment: the Asian experience. Water Sci. Technol. 31(10), 41-59.

Wahab, M.A., Azim, M.E., Mahmud, A.A., Kohinoo,r A.H.M. and Haque, M.M. 2001. Optimisation of stocking density of Thai silver barb (*Barbodes gonionotus* Bleeker) in the duckweed-fed four species polyculture system. Bangladesh J. Fish. Res. 5(1), 13-21

Zaher, M., Begum, N.N., Hoo, M.E., Begum, M. and Bhuiyan, A.K.M.A. 1995. Suitability of duckweed, *Lemna minor* as an ingredient in the feed of tilapia, *Oreochromis niloticus*. Bangladesh J. Zool. 23(1), 7-12

Chapter Four

Suitability of using duckweed as feed and treated sewage as water source in tilapia aquaculture

Submitted to Bioresource Technology as:

Saber A. El-Shafai, Huub J.Gijzen, Fayza A. Nasr, N. Peter van der Steen and Fatma A. El-Gohary. Suitability of using duckweed as feed and treated sewage as water source in tilapia aquaculture.

Suitability of using duckweed as feed and treated sewage as water source in tilapia aquaculture

Abstract

Feasibility of using both treated effluent and duckweed biomass from a pilot-scale UASB-duckweed ponds system treating domestic sewage was evaluated in rearing Nile tilapia (*Oreochromis niloticus*). The nutritional value of duckweed was compared with wheat bran, used as a local fish feed ingredient, by applying these as the only source of feed for tilapia juveniles weighing 20 grams initial mean body weight. Two sources of water were used for each feed trial, treated-sewage and freshwater. The experiment was conducted in parallel with a conventional settled sewage-fed fishpond stocked with tilapia. Results of growth performance demonstrated that, in case of freshwater ponds specific growth rate (SGR) of tilapia fed on fresh duckweed was significantly ($p < 0.01$) higher than the SGR in wheat bran fed pond. No significant difference ($p > 0.05$) was observed between the two feeding regimes in case of treated sewage-fed ponds. The SGR of tilapia reared in the treated sewage-wheat bran-fed pond (TWP) was significantly higher ($p < 0.01$) than the SGR in the freshwater-wheat bran-fed pond (FWP). On the other hand, due to the early spawning in the treated sewage-duckweed-fed pond (TDP) SGR of tilapia in the latter was significantly lower ($p < 0.05$) than the SGR in the freshwater-duckweed-fed pond (FDP). The results demonstrated that duckweed-fed ponds provide higher net fish yield (11.8 ton/ha/y in TDP and 9.6 ton/ha/y in FDP), than wheat bran-fed ponds (8.9 ton/ha/y in TWP and 6.4 ton/ha/y in FWP). The fish yields in the ponds fed with treated sewage were higher than those in the ponds fed with freshwater. Negative net yield was observed in the settled sewage-fed pond (SSP) at -0.16 ton/ha/y. The negative results obtained in this pond, were attributed to the high mortality of 60% in the adult fish and 38% in the fry during the autumn. The best result was obtained in the treated effluent-duckweed-fed fishpond, providing 11.8 ton/ha/y net yield, which shows the potential value of both treated sewage and duckweed in aquaculture of tilapia.

Keywords: Fresh duckweed; Domestic sewage; *Oreochromis niloticus*; UASB-duckweed ponds

1. Introduction

Because of the increasing demand for animal protein and limitations in the capacity of capture fisheries the importance of aquaculture as a source of fisheries products is growing. There are currently many attempts to increase fish production at all levels, notably by the efforts of the Food and Agriculture Organisation (FAO) and other national, regional and international organisations. Developing countries increasingly direct their efforts towards fish aquaculture to provide people with sufficient amounts of animal protein, essential for healthy growth.

Production of fish in developing regions is often limited by availability of nutrients in the form of fertilisers or protein-containing fish feeds. Recycling of wasted organic matter in fish aquaculture represents an attractive alternative. Reuse of wastes (organic wastes, animal wastes, agriculture wastes and human wastes) in fish production could safeguard the environment from pollution and simultaneously generate valuable marketable biomass. Availability, cost and ease of transport are important considerations when selecting a nutrient source for fish cultivation. Chemical fertilisers, though widely used in developed countries, are limited in supply and expensive in many developing countries. One of the major constraints to increase fish production in developing countries is the cost of feed and chemical fertilisers. Integration of fish and livestock production is one option to decrease fish farming cost and to increase productivity of the system (Java and Chakrabarti, 1997; Sadek and Moreau, 1998). Organic fertilisation through application of livestock manure is widely applied in fish farming (Prinsloo and Schoonbee, 1987a; Prinsloo and Schoonbee, 1987b; Zhang *et al.*, 1987; Mishra *et al.*, 1988; Green *et al.*, 1989; Edwards *et al.*, 1994). Also the use of organic manure for production of small crustaceans was carried out (Java and Chakrabarti, 1997). This biomass can subsequently be used as live food in rearing carnivorous fish. Incorporation of cheaper ingredients, like organic agro-industrial by-products and wild plants in fish feed, are options to reduce the price of fish feed and consequently fish production cost.

Availability of local ingredients for supplementary fish feed is generally limited in Africa including Egypt because of scarce agricultural production or because of strong competition with other livestock production. Egypt has to import two-thirds of its wheat and vegetable oil and currently the annual imported agricultural products amount to about 15 million ton. The nutrition gap between domestically produced food and national consumption in Egypt, which must be filled by imported food, is estimated at 40% (Abdel Mageed, 1994). Fishponds fed with conventional feed pellets containing fishmeal as a protein source result in net consumption of protein and cannot alleviate protein malnutrition in developing countries. Locally available agricultural products by-products and wild vegetation including aquatic plants can be efficiently used as sole feed input or as supplementary feed in the fertilised pond (Rifai, 1980; Cruz and Laundencia, 1980; Prinsloo and Schoonbee, 1987c; Zaher *et al.*, 1995).

The competing uses of surface water are largely restricted to irrigation systems and drinking water supply. Though the use of water in aquaculture is often considered as "non-consumptive" there are significant losses due to seepage and evaporation, depending on the soil properties and climatic conditions. In case of limitations of surface water, commercial feed and/or costly chemical fertilisers, domestic sewage could substitute both surface water and artificial pond fertilisers. The use of domestic sewage as water source and pond fertiliser is practised in a number of countries (Kutty, 1980; Edwards and Sinchumpasak, 1981; Wang, 1991) using experimental scale and full-scale ponds. Use of treated sewage in rearing local fish species was conducted in Egypt using laboratory scale models and pilot scale models (Easa *et al.*, 1995; El-Gohary *et al.*, 1995; Shereif *et al.*, 1995; Khalil and Hussein, 1997).

In fact there are two problems related to the consumption of fish raised on human wastes; the social acceptability or consumer behaviour (Mancy *et al.*, 2000) and the potential transfer of diseases. There appears to be definite cultural differences concerning the consumption of fish reared on human wastes. The Chinese and Indians appear to have few objections to eat such fish (Edwards, 1980). In the Arab world such suggestion is generally rejected. There will be a big difference in acceptability between fish directly fed on human wastes and fish fed on aquatic plant biomass grown in pre-treated sewage. The main objective of this study is the evaluation of duckweed grown on partially treated sewage as supplementary fish feed compared to wheat bran. The feasibility of using treated sewage from a UASB-duckweed ponds system as replacement of fresh water in fish aquaculture was evaluated as well.

2. Material and methods

2.1. Fish species and husbandry

Nile Tilapia (*Oreochromis niloticus*) weighing about 20 grams average weight and from the same parental stock was obtained from a hatchery two weeks before start of the experiment. The juveniles were adapted to the experimental conditions in two plastic holding tanks (48 cm deep, area 1 m^2).

The tanks were fed with de-chlorinated city tap water. Removal of chlorine was performed by addition of sodium thiosulfate solution (0.025M, 2 ml/litre) and overnight continuous aeration. Tilapia was stocked at 50 fish per tank under continuous aeration and regular removal of faeces. The water flow was 50 litres per day (9.6 days HRT). The fish was fed on a mixture of wheat bran and fresh duckweed to satiation. The experiment was conducted outdoors in five identical plastic holding tanks, each with one square meter surface area and 48 cm depth.

2.2. Water sources and fish feed

Three water sources with different qualities were used: (1) de-chlorinated city tap water, (2) treated domestic sewage from a UASB-duckweed ponds system and (3) 2 hours settled raw sewage. Two sources of supplementary fish feed were used, fresh duckweed grown on pre-treated sewage and local wheat bran.

2.3. Experimental design and operating conditions

The experiment lasted 150 days and was started by filling all tanks with the selected water sources. Pond operating conditions are shown in Table 1. Pond 1 (treated sewage-duckweed-fed pond TDP) and 2 (treated sewage-wheat bran-fed pond TWP) received treated domestic sewage at an average flow rate of 42.75 l d^{-1}. Pond 3 (freshwater-duckweed-fed pond FDP) and 4 (freshwater-wheat bran-fed pond FWP) received de-chlorinated city tap water with daily flow rate of 37.5 litre for each. Pond 5 (settled sewage-fed pond SSP) received 2 hours settled sewage at flow rate of 15 l d^{-1}. This flow rate was calculated based on the nitrogen content of the settled sewage to provide about 4 kgN/ha/d (Mara *et al.* 1993). The total nitrogen content of each pond was determined before stocking with tilapia. All ponds were stocked with tilapia juveniles at a density of 10 fish per pond. TDP and FDP were fed on fresh duckweed harvested daily from a pilot-scale UASB-duckweed ponds system treating domestic sewage. The feeding rate was established at 25 gram of fresh duckweed per 100 g of live body weight of fish. TWP and FWP were fed on wheat bran at a feeding rate providing the same amount of dry matter as applied to the duckweed-fed ponds. During the experiment fish was weighed twice, after 60 days and at the end of the experiment.

Table 1: Operating conditions in fishponds

Item	TDP	TWP	FDP	FWP	SSP
Fish weight (g fish^{-1})	21.41±2.41	20.15±3.57	19.8±3.98	21.7±2.54	20.35±2.69
Feed source	F* duckweed	Wheat bran	F* duckweed	Wheat bran	-
Feeding rate (g feed/100 g fish)	25	1.34-1.52	25	1.34-1.52	-
Water source	T* sewage	T* sewage	Freshwater	Freshwater	S* sewage
Flow rate (l d^{-1})	42.75	42.75	37.5	37.5	15

* F fresh, T treated and S settled

2.4. Studied parameters and analytical procedures

Extensive analysis of water quality in fish tanks was conducted. The measurements included water temperature, pH, dissolved oxygen (DO), chemical oxygen demand (COD), biological oxygen demand (BOD), total ammonia nitrogen, nitrite nitrogen, nitrate nitrogen, TKN, total phosphorus and total suspended solids (TSS). All the analyses were performed using the Standard Method for the Examination of Water and Wastewater, 20th edition (APHA 1998).

The un-ionised ammonia nitrogen (UIA-N) concentrations were calculated using the general equation of bases (Albert, 1973).

$$NH_3 = \frac{[NH_3 + NH_4^+]}{[1 + 10^{(pKa-pH)}]} \quad (1)$$

In fresh water the calculation of pKa is based on the equation developed by Emerson *et al* (1975).

$$PKa = 0.09018 + 2729.92/T \quad (2)$$

Where T = degree Kelvin

Estimation of nitrogen mass balance was performed through the experimental ponds by measuring the total nitrogen input, total nitrogen discharged with the effluent, total nitrogen incorporated in fish and total nitrogen in the sediment. Water discharge rate was calculated from the influent flow rate and water evaporation rate according to: Water discharge rate (l tank^{-1} d^{-1}) = Influent flow rate (l tank^{-1} d^{-1}) - Water evaporation rate (l tank^{-1} d^{-1}). The water evaporation rate (7 l tank^{-1} d^{-1}) was measured using plastic buckets. By the end of the experiment, the accumulated sediments in the ponds were analysed for N. Duckweed and wheat bran were analysed for dry matter and nutrients (N and P) content. Protein content was calculated from the organic nitrogen where protein = N × 6.25 (Rusoff *et al.*, 1980). The dry matter content was calculated after drying at 70 °C overnight. Organic nitrogen was analysed using Kjeldahl method while phosphorus content was determined using vanado-molybdate method following per-sulphate digestion method. The specific growth rates were calculated individually for each treatment based on the following expression:

$$SGR = \frac{(\ln Wf - \ln Wi)}{days} \times 100 \qquad (3)$$

Where, Wi and Wf are initial and final mean body weight respectively.

The total feed input as dry matter was recorded in each treatment, and feed conversion ratio (FCR) was calculated based on the following expression:

$$FCR = \frac{total\ feed\ ingested\ (dry\ weight)}{fish\ weight\ gain\ (wet\ weight)} \qquad (4)$$

$$Protein\ efficiency\ ratio\ (PER) = \frac{fish\ weight\ gain}{total\ protein\ applied} \qquad (5)$$

By the end of the trial, fish mortalities, total fish yield and net fish yield in ton/ha/year and nitrogen mass balance were calculated for each treatment.

$$Total\ fish\ yield = \frac{(Final\ density\ in\ ton\ /\ ha)(365)}{Growth\ period\ in\ days} \qquad (6)$$

$$Net\ fish\ yield = \frac{(Initial\ density\ in\ ton\ /\ ha - Final\ density\ in\ ton\ /\ ha)(365)}{Growth\ period\ in\ days} \qquad (7)$$

2.5. Statistical analysis

The parallel treatments were subjected to one-way analysis of variance (one-way ANOVA) to investigate the significance of differences.

3. Results

3.1. Water quality and fish feed

The water quality of treated effluent and settled sewage is presented in Table 2. Monitoring the fishponds has been performed over the experimental period. The results are presented in Table 3. No accumulation of ammonia or nitrite occurred in the ponds except for the pond fed with settled sewage. The protein and phosphorus content of duckweed and wheat bran are presented in Table 4.

Table 2: Characteristics of treated and settled sewage used in the experiment

Parameters	Treated sewage	Settled sewage
Temperature (^{o}C)	24-34	25-31
pH	7.2-8.3	6.6–7.5
COD (mg O_2 l^{-1})	49±18	521±136
BOD (mg O_2 l^{-1})	14±5	208±46
TAN* (mg N l^{-1})	0.4±0.9	17.9±3.3
Nitrite(mg N l^{-1})	0.166±0.104	0.01±0.02
Nitrate (mg N l^{-1})	0.697±0.676	0.13± 0.03
TKN (mg N l^{-1})	4.4±1.4	23.2±5.4
TP (mg P l^{-1}l)	1.11±0.41	3.58±0.84
TSS (mg l^{-1})	32±10	161± 74
DO (mg O_2 l^{-1})	6.3±1.2	Nil

*TAN total ammonia nitrogen

Table 3: Water quality parameters in the fishponds

Parameters	TDP	TWP	FDP	FWP	SSP
Temperature ^{o}C	23-35	23-35	23-35	23-35	23-35
pH	8-9.4	8-9.7	8.4-9.9	7.4-9	8.1-10.7
COD (mg O_2 l^{-1})	295±104	257±68	108±46	110±65	256±61
BOD (mg O_2 l^{-1})	34±11	28±8	18±9	19±10	28±5
TAN (mg N l^{-1})*	0.22±0.29^a	0.15±0.2^a	0.12±0.2^{ab}	0.05±0.08^b	0.57±0.89^c
UIA-N (mg N/l)*	0.06±0.12^a	0.04±0.08^a	0.05±0.11^{ab}	0.004±0.01^b	0.12±0.22^c
Nitrite (mg N l^{-1})**	0.024±0.012^a	0.02±0.016^a	0.02±0.01^a	0.01±0.005^a	0.11±0.15^b
Nitrate (mg N l^{-1})	0.181±0.07	0.132±0.06	0.1±0.055	0.09±0.04	0.138±0.055
TKN (mg N l^{-1})	5.95±1.58	4.6±1.02	1.41±0.38	0.4±0.14	5.92±1.45
TP (mg P l^{-1})	1.09±0.43	0.88±0.36	0.56±0.19	0.32±0.11	1.4±0.57
TSS (mg l^{-1})	195± 92	164±59	82±48	83±48	135±34
DO (mg O_2 l^{-1})	15.9± 4.7	14.9±4.7	15.3± 4.1	13.8±4.7	27.3±8.1

*$P<0.05$

**$P<0.01$

Statistical comparison calculated between values within the same row. Values with the same superscript letters are not significantly different

Table4: Quality of duckweed and wheat bran

Parameters	Duckweed	Wheat bran
Dry matter (%)	5.2±0.5	90.7±2.1
Protein content (%)	21.1± 0.51	11±1.3
Phosphorus content (%)	0.69±0.3	0.43±0.02

Table 5: Growth performance of tilapia (*Oreochromis niloticus*) in fishponds

Parameters	TDP	TWP	FDP	FWP	SSP
Initial standing crop density (g tank^{-1})	214.1	201.5	198	217	203.5
Final standing crop density (g tank^{-1})	481.5	448.9	477.5	432.5	157
Weight gain (g tank^{-1})	267.4	247.4	279.5	215.5	-
Feed input (g tank^{-1})[**]	589	585	561	526.5	-
SGR[*]	0.53±0.06^{a*}	0.53±0.06^{a*}	0.59±0.05^{b*}	0.46± 0.04^{c}	0.44
FCR	2.2	2.36	2.01	2.44	-
PER	2.1	3.84	2.12	3.72	-

[*] $p < 0.05$ _ the underscore means that $p < 0.01$ in significantly different treatments (e.g., TDP-FWP $P< 0.05$ and TWP-FWP $P< 0.01$),

[**] Feed input in dray matter

3.2. Growth performance

No significant difference ($p > 0.05$) between the SGR of tilapia stocked in treated effluent and fed on cither fresh duckweed or wheat bran was detected (Table 5). On the other hand, when using freshwater, the juvenile tilapia fed on fresh duckweed had significant higher ($p < 0.01$) SGR than those fed on wheat bran. Significantly, lower SGR ($p < 0.05$) was observed for duckweed-fed tilapia when using treated sewage instead of freshwater. When using wheat bran as a feed the opposite effect was observed. SGR in treated sewage-fed pond was significantly ($p < 0.01$) higher compared to the freshwater-fed pond. Due to high mortality in the settled sewage-fed pond, statistical analysis was not performed for this pond and the SGR was calculated based on the average fish weight, initial and final.

Total and net fish yield was calculated in all ponds and presented in Figure 1. At the end of the trial, a relatively large part of the total fish biomass consisted of fry fish, especially in TDP (Fig. 2). The total net fish yields (fry and adult) were 11.8, 8.9, 9.6, 6.4 and −0.16 ton/ha/y in the TDP, TWP, FDP, FWP and SSP, respectively. The fry fish contributed to the net yield by 45%, 32%, 29% and 18% in TDP, TWP, FDP and FWP, respectively.

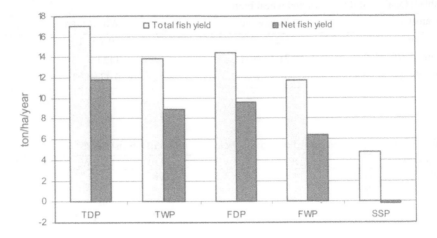

Fig.1: Total and net fish yield in fishponds

3.3. Nitrogen mass balance

The results shown in Table 6 show that the total nitrogen recovery in fish was 19.6% and 17.9% from the total nitrogen input (feed + treated sewage) in the TDP and TWP, respectively. The values in freshwater fed ponds were 40% and 57.4% in FDP and FWP, respectively. The total nitrogen recovery in SSP was estimated to be 10% including died fish.

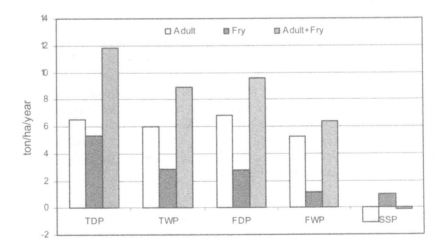

Fig. 2: Net fish yield in fishponds

Table 6: Nitrogen mass balance in duckweed, wheat bran and settled sewage fishponds

Item	TDP	TWP	FDP	FWP	SSP
$TN_{intput.}$	54.18	44.07	21.05	9.27	52.2
$TN_{reco.}$	10.61 (19.6%)	7.88 (17.9%)	8.4 (40%)	5.32 (57.4%)	5.29 (10%)
$TN_{disc.}$	36.7 (67.7%)	31.57 (71.6%)	7.8 (37%)	2.58 (27.8%)	9.35 (18%)
$TN_{sed.}$	6.5 (12%)	4.2 (9.5%)	3.8 (18%)	0.6 (6.5%)	11.8 (23%)
$TN_{unacc.}$	0.37 (0.7%)	0.42 (1%)	1.05 (5%)	0.77 (8.3%)	25.76 (49%)

[*]$TN_{input.}$ total nitrogen input, $TN_{reco.}$ total nitrogen recovered, $TN_{disc.}$ total nitrogen discharged, $TN_{sed.}$ Total nitrogen in the sediment and $TN_{unacc.}$ Unaccounted for nitrogen (all in gram),
[**] Values between parentheses are % of total nitrogen input.

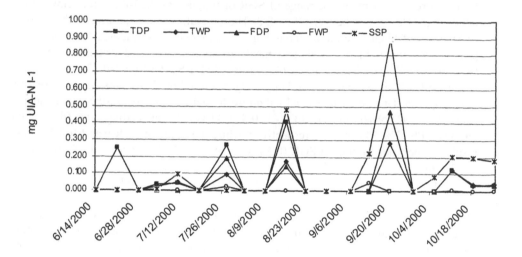

Fig. 3: UIA-N concentrations in fishponds in mg UIA-N/l

4. Discussion

4.1. Water quality and fish mortality

Significant higher values of ammonia ($p < 0.05$) and nitrite ($p < 0.01$) were detected in the SSP due to the accumulation of ammonia and low nitrification at the end of the experiment (Autumn). This was mainly due to partial decrease in temperature and decrease in day length. This lead to decreased photosynthetic activity and ammonia uptake capacity of the phytoplankton in SSP. Fish mortality in this pond was 60% in the adult fish and 38% in the fry. This mortality was probably due to chronic ammonia toxicity rather than due to an acute toxicity effect. Acute ammonia toxicity for tilapia may occur at values above 2 mg UIA-N l^{-1} (Render and Stickeny, 1979) while the maximum un-ionised ammonia concentration recorded in SSP was 0.9 mg UIA-N l^{-1} (Fig. 3). Person-Le Ruyet et al. (1997) demonstrated, 50% mortality in trout juvenile after prolonged exposure (chronic toxicity) to 0.73 mg UIA-N l^{-1}. In fresh duckweed-fed tilapia no mortality was recorded after 75 days exposure to 0.43 mg UIA-N l^{-1} (El-Shafai *et al.*, in press). However, Nasr *et al.* (1998) reported skin ulcer, necrosis and mortality in tilapia after 30 days stocking period in treated sewage pond at 0.45 mg UIA-N l^{-1}. Nitrite toxicity also may have played a role in the toxicity either directly (Anuradha and Subburam, 1995; Woo and Chiu, 1995; Schoore *et al.*, 1995) by oxidising haemoglobin into non-functional methaemoglobin and/or indirectly (Bunch and Bijerano, 1997) by increasing susceptibility of tilapia for bacterial infection.

4.2. Growth performance and fish yield

In the current experiment, the range of SGR of tilapia was 0.53-0.59 in duckweed-fed ponds and FCR ranged between 2.01-2.2. These results are comparable with the SGR (0.51-0.73) of tilapia weighing about 5 grams and fed on formulated feed with 50% protein, containing blood meal, fish meal, groundnut cake and water hyacinth leaf (Keke *et al.*, 1997). Gaigher *et al.* (1984) reported a SGR of 0.63-0.72 when rearing tilapia with 2.8 grams mean initial body weight and fed on fresh duckweed. By rearing Nile tilapia in static water tanks and fed on fresh duckweed (*Lemna gibba*), Hassan and Edwards (1992) observed SGR of 0.41, 0.8, 1.34 and 1.4 at feeding rates (dry matter bases) of 1%, 2%, 3% and 4%, respectively while the range of FCR was 1.9-3.3. These results support our finding of SGR (0.53-0.59) at 25% feeding rate on fresh duckweed (1.13% dry matter bases). The FCR of 1.9 observed by Hassan and Edwards (1992) at 1% feeding rate is similar to our results (2.01-2.2) in duckweed fed ponds.

SGR in our experiment is lower than reported values (1.8-2.2) of tilapia with 2.5 grams mean body weight and fed on three isonitrogenous fishmeal-based diets containing local ingredients (water hyacinth, duckweed and rice bran) (Zaher *et al.*, 1995). The FCR (2.2-2.5) of these three diets is similar to our data. The low SGR in our study is mainly attributed to the low feeding rate rather than to the quality of the duckweed. Zaher *et al.* (1995) applied a feeding rate of 3-5% (dry matter bases) that is 3-4 times higher than in this study.

Our results of FCR and PER are better than the reported values (4.3 and 0.8, respectively) for tilapia (13.5 grams) fed on formulated feed with sun-dried *Spirodela* as substitute for fishmeal (Fasakin *et al.*, 1999), which suggests a better nutritional value of *Lemna gibba* in comparison to *Spirodela*. Our results are comparable with the results of FCR (1.2-2) and better than the results of PER (1.1-1.8) reported by Fontainhas *et al.* (1999) for tilapia fed on fishmeal-based diets with 0%, 33%, 66% and 100% plant ingredients (extruded pea seed meal, defatted soybean meal, full-fat toasted soybean and micronized wheat) as replacement of fishmeal.

Our results showed higher SGR ($p < 0.01$) of tilapia in FDP than in FWP. This is probably due to the high nutritional value of duckweed (high protein content) over the wheat bran, Table 4. The dietary protein is of fundamental importance and generally represents the limiting factor, determining growth performance in fish aquaculture systems. In case of wheat bran-fed ponds, significantly higher SGR was observed in TWP ($p < 0.01$) than in the FWP. The nutrients in the treated sewage increased the natural productivity of the pond. The natural productivity of the pond significantly contributes to satisfy the demand for essential compounds such as vitamins, cholesterol, phosphorus and minerals (D' Abramo and Conklin, 1995). Also Cruz and Laudencia (1980) reported that the net fish production and growth performance of fish was significantly higher in fertilised ponds fed on rice bran or copra meal than the yield in non-supplemented ponds. Agustin (1999) investigated the effect of natural productivity and formulated feed addition on the growth of fresh water prawn, *Macrobrachium borelli*. He reported that the natural productivity supported 43% of the prawn growth. In supplemented earthen ponds, the growth of freshwater prawn (*Macrobrachium rosenbergi*) was significantly higher in organic fertilised ponds than in the non-fertilised ponds (Tidwell *et al.*, 1995). The role of natural productivity in enhancing fish yield was also reported in tilapia fertilised ponds (cow manure) fed on formulated feed, in comparison to non-fertilised ponds fed on formulated feed (Victor, 1993).

In addition to the treated sewage nutrients, considerable amount of daphnia and copepods were observed in the treated effluent, Figure 4. The published data show that the protein content of daphnia grown on wastewater is in the range of 55-65% (dry matter) while the energy content is representing 20 KJ g^{-1} (Proulx and De la Noue, 1985; Kibria *et al.*, 1999). The daphnia and copepods contain high amounts of essential and non-essential fatty acids (McEvoy *et al.*, 1998; Kibria *et al.*, 1999). They represent considerable part of the prey for the omnivorous and carnivorous fish in the natural water and organic fertilised fishponds (Warburton *et al.*, 1998; Kibria *et al.*, 1999; Lienesch and Gophen, 2001).

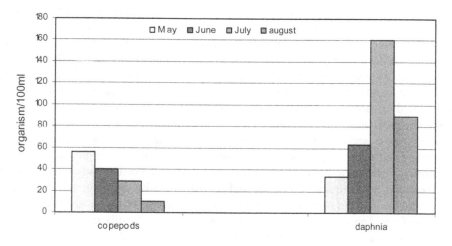

Fig. 4: Count of copepod and daphnia in the treated sewage used in TDP and TWP

Absence of any significant differences between the TDP and TWP and significantly higher SGR in case of FDP than in the TDP is attributed to the early spawning that occurred in the TDP in comparison to the other ponds, Table 7. The new larvae increased the stocking density and negatively affected the SGR and FCR of the adult through competition for feed (Canario *et al.*, 1998) and enhancing social interaction due to high density (Breine *et al.*, 1996). The spawning itself consumed large amount of energy and so decreased the energy budget of the fish with negative effects on the growth performance. The dietary protein is known to positively affect spawning in fishponds (Siddiqui *et al.*, 1998). Accordingly the early spawn in the TDP could be attributed to high protein input either directly from the duckweed, daphnia and copepods or indirectly from the pond fertilisation by the treated sewage.

Table 7: Production of fry in fishponds

Fish ponds Parameters	TDP		TWP		FDP	FWP	SSP
No of fry (fry tank^{-1})	4	58	73	83	60	28	34
Average weight (g fish^{-1})	10.9	2.99	1.21	0.38	1.93	1.68	1.9
Age (days)	125	50	50	35	50	50	125
No of fry female^{-1}	31		52		20	14	11
No of spawn femal^{-1}	2		2		1	1	1
% of mortality	0		0		0	0	38

The total net fish yields, including adult fish and fry, were 11.8 and 9.6 ton ha^{-1} y^{-1} in the TDP and FDP, respectively. In the parallel treatments of wheat bran the net yields were 8.9 and 6.4 ton ha^{-1} y^{-1} in TWP and FWP respectively. The values for duckweed as well as the values for the treated sewage-fertilised ponds show the advantages of duckweed over the wheat bran and the role of treated sewage in enhancing the fish yield. In organic (cow manure) fertilised ponds supplemented with formulated feed with 27% protein, the net yield of tilapia reached 11.8 ton ha^{-1} y^{-1} (Victor, 1993). This yield is similar to the yield of TDP in our trial, which confirms that the nutritional value of duckweed and treated sewage is similar to that of commercial feed containing 27% protein and cow manure. Production of polyculture carp fed with fresh duckweed and treated sewage ranged between 10-15 ton ha^{-1} y^{-1} (Skillicorn *et al.*, 1993) while the maximum attainable yield in ponds fed with fresh duckweed and other supplementary feeds (rice bran and oil cake) was 6.3 ton ha^{-1} y^{-1} (Wahab *et al.*, 2001). The lower yield in sewage-fed ponds supplemented with rice bran and oil cake might be attributed to the higher fibre and lower protein content of rice bran and/or the bad effect of oil cake on the pond water quality. The oil and fat of oil cake disperse in the pond and contribute to the BOD and decrease the gas transfer by forming a fatty layer on the pond surface. The yearly production of tilapia in ponds fertilised with chicken litter was 4.3 ton ha^{-1} y^{-1} (Green *et al.*, 1989). The results of duckweed-fed ponds and treated sewage-fed ponds is higher than the reported values for the sewage-fed ponds (Shereif *et al.*, 1995), buffalo manure-fed ponds (Edwards *et al.*, 1994) and swine manure-fed ponds (Berends *et al.*, 1980). This indicating the good quality of fresh duckweed as supplemental feed for sewage fed ponds or organic manure fed ponds. This also shows the advantage of producing duckweed on sewage as fish feed and its use with the treated effluent in tilapia farming instead of using the sewage directly as source of pond fertilisation. On the other hand this data is lower than the 20 ton ha^{-1} yearly production rate of polyculture carp in an integrated duck-fish pond fed with formulated feed in addition to duck faeces (Prinsloo and Schonbee, 1987a). This might be attributed to the use of a carp polyculture, which covers all the trophic levels in the pond and therefore decreases feed loss.

4.3. Nitrogen mass balance

18% of the total nitrogen input (dietary N + N in the influent) in the treated sewage fed ponds was recovered in fish while the percentage was 40% and 57% in the FDP and FWP, respectively. The values in fertilised ponds supplied with supplementary feed (duckweed or wheat bran) are within the range of 5-25% reported for semi-intensive fishpond culture (Tacon et al., 1995). The low value of 10% N-recovery obtained in SSP is still higher than the 3.5% that was reported in polyculture fishponds with high mortality (Liang et al., 1999). In the fresh water ponds the range (40-57%) of N recovery is similar to 53% reported by Gaigher et al. (1984) for tilapia reared in re-circulating tanks and fed on fresh duckweed. The N retained in the sediment was in the range of 3.8-11.8%, which is similar to the range of 4.3-13.7% reported by Gomes et al. (1995) in trout growth trials using plant-based formulated diets. Gaigher et al. (1984) reported that 14% of the N from fresh duckweed accumulated in the sediment.

Two primary processes affect ammonia concentration, i.e. fish excretion and sediment diffusion (Hargreaves, 1997). The amount of ammonia excreted by fish depends on average feeding rate and dietary protein content and its biodegradability. In semi-intensive aquaculture systems, nitrogen dynamics in the water column are controlled by ammonia input, its assimilation by phytoplankton or nitrification and loss of nitrogen through sedimentation, volatilisation and discharge. Seawright et al. (1998) highly-lighted the role of phytoplankton in nitrogen dynamics in fishponds. This is supported by Lorenzen et al. (1997) who reported that assimilation by phytoplankton and subsequent sedimentation is the principal process of ammonia removal. In the current experiment the unaccounted for part of nitrogen, which was probably lost via ammonia volatilisation and/or nitrogen denitrification, was small in treated sewage fed ponds. The major source of ammonia in these ponds was probably ammonia excretion by fish, and this remained below the phytoplankton uptake capacity. The higher value of unaccounted for nitrogen in freshwater fed ponds could be attributed to the low density of algae in these ponds. When the inputs of ammonia exceed the phytoplankton assimilation capacity, the remaining high ammonia concentration is subject to volatilisation and de-nitrification. This is what happened in the SSP that received considerable amounts of ammonia nitrogen in the settled sewage, Table 2. The high pH value in the SSP pond, which maximally reached 10.7, enhanced ammonia volatilisation (Gomez et al., 1995). Dissolved oxygen produced as by-product from photosynthesis might enhance the nitrification process (Liang et al., 1998). The produced nitrate serves as substrate for denitrification during the night. During the night, low dissolved oxygen concentration due to absence of photosynthesis and presence of respiration by algae and fish and presence of high organic matter in SSP may have enhanced the de-nitrification process (Daniels and Boyd, 1989; Diab et al., 1993).

4.4. Nitrogen wastes

The total nitrogen waste in the ponds ranged between 12.1- 97.8 Kg N/ton fish production, Figure 6. The range for the treated sewage-fed ponds was 89.1-97.8 Kg N ton fish[-1], which is comparable to the amount of nitrogen released from production of one ton of carp (53-71 kg N) and the nitrogen waste (26-117 Kg N) for one ton of shrimp production (Tacon et al., 1995). In typical Nordic freshwater fish farms the nitrogen release to the environment was estimated to be 132 kg N in 1974 and 55 Kg N in 1995 for each ton of fish production (Enell, 1995) while Gomez *et al.* (1995) reported 61-129 kg N/ton fish in rainbow trout. The nitrogen waste in our freshwater-fed ponds (12.1-29.4 Kg N/ton fish) is low, which is mainly attributed to the low amount of nitrogen applied to the ponds. The nitrogen solid waste (faecal N in the sediment) in our experiment ranged between 5-9 Kg N for each ton of fish production. Gomes *et al.* (1995) showed that the solid waste of N released from a rainbow trout farm ranged between 6.3 and 16.1 Kg N/ton fish.

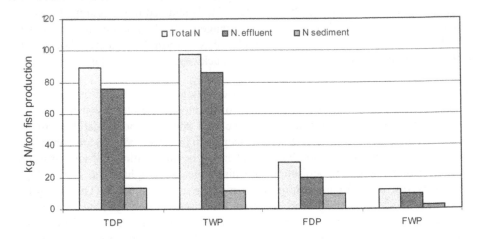

Fig. 5: Nitrogen wastes in fishponds

5. Conclusions

-The experiments have shown that it is feasible to use treated domestic sewage from UASB-duckweed pond systems as replacement of freshwater in tilapia farming.
-Fresh duckweed (*Lemna gibba*) is superior to wheat bran as supplementary feed in tilapia pond farming. Main advantages of duckweed are the high protein content and its low cost.
-Fertilisation of tilapia ponds with treated sewage from UASB-duckweed pond increases the fish yield significantly. This was valid for both ponds fed with duckweed and wheat bran.
-Since the most important water quality parameters in fishponds are ammonia, nitrite and DO, it is recommended to cultivate duckweed in anaerobically treated sewage followed by use of both treated sewage and duckweed biomass in tilapia farming instead of direct use of settled sewage as a pond fertiliser.

Acknowledgements

The authors would like to thank the Dutch government for financial support for this research within the framework of the SAIL funded "Wasteval" project (LUW/MEA/971). The project is a co-operation between the Water Pollution Control

Department (NRC, Egypt), Wageningen University and UNESCO-IHE Institute for Water Education, The Netherlands. The authors would like to express their sincere thanks to the technicians of Water Pollution Control Department, NRC, Egypt

References

Abdel Mageed, Y. 1994. Water in the Arab world, perspectives and prognoses, the central region: problems and perspective. In: Peter Rogers and Peter Lydon (Eds.), Water in the Arab world, perspectives and prognoses, pp. 101-119

Agustin, C.P. 1999. Role of natural productivity and artificial feed in the growth of freshwater prawn, *Macrobrachium borelli* (Nobili, 1896) cultured in enclosures. J. Aquacult. Trop., 14(1), 47-56

Albert, A. 1973. Selective toxicity. Chapman and Hall, London.

American Public Health Association 1998. Standard Methods for the Examination of Water and Wastewater, 20th edition, Washington.

Anuradha, S. and Subburam, V. 1995. Role of sewage bacteria in methemoglobin formation in *Cyprinus carpio* exposed to nitrate. J. Environ. Biol. 16(2), 175-179

Behrends, L.L., Maddox, J.J., Madewell, C.E. and Pile, R.S. 1980. Comparison of two methods of using liquid swine manure as an organic fertiliser in the production of filter-feeding fish. Aquaculture 20, 147-153.

Bunch, E.C. and Bejerano, I. 1997. The effect of environmental factors on the susceptibility of hybrid tilapia *Oreochromis niloticus* X *Oreochromis aureus* to streptoccosis. Bamidgeh 49(2), 67-76.

Breine, J. J., Nguenga, D., Teugels, G.G. and Ollevier, F. 1996. A comparative study on the effect of stocking density and feeding regime on growth rate of *Tilapia cameronensis* and *Oreochromis niloticus* (Cichlidae) in fish culture in Cameroon. Aquat. Living Resour. 9, 51-56.

Canario, A.V.M., Condeca, J. and Power, D.M. 1998. The effect of stocking density on growth in the gilthead sea bream, *Sparus aurata* (L.). Aquacult. Res. 29, 177-181.

Colt, J., Ludwig, R., Tchobanoglous, G. and Cech, J. 1981. The effects of nitrite on the short-term growth and survival of channel catfish, *Ictalurus punctatus*. Aquaculture 24, 111-122.

Cruz, E.M. and Laundencia, I.L. 1980. Polyculture of milkfish (*Chanos chanos* Forskal), all-male Nile tilapia (*Tilapia nilotica*) and snakehead (*Ophicephalus striatus*) in freshwater ponds with supplemental feeding. Aquaculture 20, 231-237.

D'Abramo, L.R. and Conklin D.E. 1995. New development in the understanding of the nutrition of penaeid and caridean species of shrimp. Proceedings special session on shrimp farming, aquaculture 95, World Aquaculture Society, Louisiana, USA, pp. 95-107

Diab, S., Kochab, M. and Avnimelech, Y. 1993. Nitrification pattern in a fluctuating anaerobic-aerobic pond environment. Water. Res. 27(9), 1469-1475.

Daniels, H.V. and Boyd, C.E. 1989. Chemical budgets for polyethylene line, brackish water ponds. J. World Aquaculture Society 20(2), 53-59

Easa, M.E., Shereif, M.M., Shaaban, A.I. and Mancy, K.H. 1995. Public health implication of wastewater reuse for fish production. Water Sci. Technol. 32(11), 145-152.

Edwards, P. 1980. A review of recycling organic wastes into fish, with emphasis on the tropics. Aquaculture 21, 261-279.

Edwards, P. and Sinchumpasak, O. 1981. The harvest of microalgae from the effluent of sewage fed high rate stabilisation pond by *Tilapia nilotica* part 1: describtion of the system and the study of the high rate pond. Aquaculture 23, 83-105.

Edwards, P., Kaewpaitoon, K., Little, D.C. and Siripandh, N. 1994. An assessment of the role of buffalo manure for pond culture of tilapia. II: Field trial. Aquaculture 126, 97-106.

El-Gohary, F., El-Hawarry, S., Badr, S. and Rashed, Y. 1995. Wastewater treatment and reuse for fish aquaculture. Water Sci. Technol. 32(11), 127-136.

El-Shafai, S.A., Gijzen, H.J., Nasr, F.A., Van Der Steen, N.P. and El-Gohary, F.A. (Submitted). Chronic ammonia toxicity to duckweed-fed tilapia (*Oreochromis niloticus*).

Emerson, K.R., Russo, R.C., Lund, R.E. and Thurston, R.V. 1975. Aquous ammonia equilibrium calculations: effect of pH and temperature. J. Fish. Res. Board Can. 32, 2377-2383

Enell, M. 1995. Environmental impact of nutrients from Nordic fish farming. Water Sci. Technol. 31(10), 61-71

Fasakin, E.A., Balogun, A.M. and Fasuru, B.E. 1999. Use of duckweed, *Spirodela polyrrhiza* L. schleiden, as a protein feed stuff in practical diets for tilapia, *Oreochromis niloticus* L. Aquacult. Res. 30, 313-318

Fontainhas, F. A., Gomes, E., Reis-Henriques, M.A. and Coimbra, J. 1999. Replacement of fishmeal by plant proteins in the diet of Nile tilapia: digestibility and growth performance. Aquacult. International 7, 57-67

Gaigher, I.G., Porath, D. and Granoth, G. 1984. Evaluation of duckweed (*Lemna gibba*) as feed for tilapia (*Oreochromis niloticus* × *Oreochromis aureus*) in a recirculating unit. Aquaculture 41, 235-244.

Gomez, E., Casellas, C., Picot, B. and Bontoux, J. 1995. Ammonia elimination processes in stabilization and high-rate algal pond systems. Water Sci. Technol. 31(12), 303-312

Green, B.W., Pheleps, R.P. and Alvarenga, H.R. 1989. The effects of manure and chemical fertilisers on the production of *Oreochromis niloticus* in earthen ponds. Aquaculture 76, 37-42.

Hargreaves, J.A. 1997. A simulation model of ammonia dynamics in commercial catfish ponds in the southeast United States. Aquacult. Eng. 16, 27-43

Java, B.B. and Chakrabarti, L. 1997. Effect of manuring rate on in situ production of zooplankton *Daphnia carinata*. Aquaculture 156, 85-99.

Keke, I.R., Ofajekwu, C.P, Ufodike, E.B. and Asala, G.N. 1994. The effect of partial substitution of groundnut cake by water hyacinth (*Eichhornia crassipes*) on growth and food utilisation in the Nile tilapia, *Oreochromis niloticus* (L.). Acta Hydrobiol. 36(2), 235-244

Khalil, M.T. and Hussein, H.A. 1997. Use of wastewater for aquaculture: an experimental field study at a sewage treatment plant, Egypt. Aquacult. Res. 28, 859-865.

Kibria, G., Nugegoda, D., Fairclough, R., Lam, P. and Bradley, A. 1999. Utilisation of wastewater-grown zooplankton: Nutritional quality of zooplankton and performance of silver perch *Bidyanus bidyanus* (Mitchell 1838) (Teraponidae) fed on wastewater-grown zooplankton. Aquacult. Nutr. 5, Pp. 221-227

Kutty, M.N. 1980. Aquaculture in South East Asia: some points of emphasis. Aquaculture 20, 159-168.

Liang, Y., Cheung, R.Y.A., Everitt, S. and Wong, M.H. 1998. Reclamation of wastewater for polyculture of freshwater fish: wastewater treatment in ponds. Water Res. 32(6), 1864-1880.

Lienessch, P.W. and Gophen, M. 2001. Predation by inland silversides on exotic cladoceran, *Daphnia lumholtzi*, in lake Texoma, USA. J. Fish Biol. 59, 1249-1257

Lorenzen, K., Struve, J. and Cowan, V.J. 1997. Impact of farming intensity and water management on nitrogen dynamics in intensive pond culture: a mathematical model applied to Thai commercial shrimp farms. Aquacult. Res. 28, 493-507

Mancy, K.H., Fattal, B. and Kelada, S. 2000. Cultural implications of wastewater reuse in fish farming in the Middle East. Water Sci. Technol. 42(1-2), 235-239.

Mara, D.D., Edwards, P., Clark, D. and Mills, S.W. 1993. A rational approach to the design of wastewater-fed fishponds. Water Res. 27(12), 1797-1799.

McEvoy, L.A., Naess, T., Bell, J.G. and Lie, Ø. 1998. Lipid and fatty acid composition of normal male-pigmented Atlantic halibut (*Hippoglossus hippoglossus*) fed enriched Artemia: A comparison with fry fed wild copepods. Aquaculture, 163, Pp. 237-250

Mishra, R.K., Sahu, A.K. and Pani, K.C. 1988. Recycling of the aquatic weed, water hyacinth and animal wastes in the rearing of Indian major carps. Aquaculture 68, 59-64.

Nasr, F.A., El-Shafai, S.A. and Abo-Hegab, S. 1998. Suitability of treated domestic wastewater for raising *oreochromis niloticus*. Egypt. J. Zool. 31, 81-94

Person-Le Ruyet, J., Delbard, C., Chartois, H. and Le Dellion, H. 1997. Toxicity of ammonia to turbot juveniles: effects on survival, growth and food utilisation. Aquat. Living Resour. 10, 307-317.

Prinsloo, J.F. and Schoonbee, H.J. 1987a. Investigations into the feasibility of a duck-fish-vegetable integrated agriculture-aquaculture system for developing areas in South Africa. Water SA 13(2), 109-118.

Prinsloo, J.F. and Schoonbee, H.J. 1987b. The use of sheep manure as nutrient with fish feed in pond fish polyculture in Transkei. Water SA 13(2), 119-123.

Prinsloo, J.F. and Schoonbee, H.J. 1987c. Growth of the Chinese grass carp *Ctenopharyngodon idella* fed on cabbage wastes and Kikuyu grass. Water SA 13(2), 125-128.

Proulx, D. and De La NoÜe, J. 1985. Harvesting *Daphnia magna* grown on urban tertiary treated effluents. Water Res. 19(10), 1319-1324.

Render, B.D. and Stickney, R.R. 1979. Acclimatisation to ammonia by tilapia aureus. Trans. Am. Fish. Soci., 108, 383-388

Rifai, S.A. 1980. Control of reproduction of *Tilapia nilotica* using cage culture. Aquaculture 20, 177-185.

Rusoff, L.L., Blakeney, E.W.Jr. and Culley, D.D.Jr. 1980. Duckweeds (*Lemnaceae Family*): A potential source of protein and amino acids. J. Agric. Food Chem. 28, 848-850.

Sadek, S. and Moreau, J. 1998. Culture of Macrobrachium rosenbergii in monoculture and polyculture with Oreochromis niloticus in paddies in Egypt. Bamidgeh 51(1), 33-42.

Schoore, J.E., Simco, B.A. and Davis, K.B. 1995. Responses of blue catfish and channel catfish to environmental nitrite. J. Aquat. Animal Health 7, 304-311

Seawright, D.E., Stickney, R.R. and Walker, R.B. 1998. Nutrient dynamics in integrated aquaculture-hydroponics systems. Aquaculture 160, 215-237

Shereif, M.M., Easa, M.E., El-Samra, M.I. and Mancy, K.H. 1995. A demonstration of wastewater treatment for reuse applications in fish production and irrigation in Suez, Egypt. Water Sci. Technol. 32(11), 137-144.

Siddiqui, A.Q., Al-Hafedh, Y.S. and Ali, S.A. 1998. Effect of dietary protein level on the reproductive performance of Nile tilapia, *Oreochromis niloticus* (L.). Aquacult. Res. 29, 349-358

Skillicorn, P., Spira, W. and Journey, W. 1993. Duckweed aquaculture " A new aquatic farming system for developing countries. The international bank for reconstruction and development / The World Bank, Washington DC.

Tacon, A.G.J., Phillips, M.J. and Barg, U.C. 1995. Aquaculture feeds and the environment: the Asian experience. Water Sci. Technol. 31(10), 41-59.

Tidwell, J.H., Webster, C.D., Sedlacek, J.D., Weston, P.A., knight, W.L., Hill, Jr S.J., D'Abramo, L.R., Daniels, W.H., Fuller, M.J. and Montanez, J.L. 1995. Effects of complete and supplemental diets and organic pond fertilisation on production of *Macrobracheium rosenbergii* and associated benthic macroinvertebrate populations. Aquaculture, 138, 169-180

Victor, P. 1993. Growth response of Nile tilapia to cow manure and supplemental feed in earthen ponds. Revue d'Hydrobiologie Tropicale 26(2), 153-160

Wahab, M.A., Azim, M.E., Mahmud, A.A., Kohinoor, A.H.M. and Haque, M.M. 2001. Optimisation of stocking density of Thai silver barb (*Barbodes gonionotus* Bleeker) in the duckweed-fed four species polyculture system. Bangladesh J. Fish. Res. 5(1), 13-21

Wang, B. 1991. Ecological waste treatment and utilisation systems on low-cost, energy-saving/ generating and resources recoverable technology for water pollution control in China. Water Sci. Technol. 24(5), 9-19.

Warburton, K., Retif, S. and Hume, D. 1998. Generalists as sequential specialists: diets prey switching in juvenile silver perch. Environmental biology of Fishers, 51, Pp. 445-454

Woo, N.Y.S. and Chiu, S.F. 1995. Effect of nitrite exposure on hematological parameters and blood respiratory properties in the sea bass *Lates calcarifer*. Environmental Toxicology and Water Quality 10, 259-266.

Zaher, M., Hoo, M.E., Begum, M.N. and Bhuyan, A.K.M.A. 1995. Suitability of duckweed, *Lemna minor* as an ingredient in the feed of tilapia, *Oreochromis niloticus*. Bangladesh J. Zool. 23(1), 7-12.

Zhang, F.L., Zhu, Y. and Zhou, X.Y. 1987. Studies on the ecological effects of varying the size of fishponds loaded with manure and feeds. Aquaculture 60, 107-116.

Chapter Five

Chronic ammonia toxicity to duckweed-fed tilapia (*Oreochromis niloticus*)

In press in Aquaculture as:

Saber A. El-Shafai, Huub J.Gijzen, Fayza A. Nasr, N. Peter van der Steen and Fatma A. El-Gohary. Chronic ammonia toxicity to duckweed-fed tilapia (*Oreochromis niloticus*).

Chronic ammonia toxicity to duckweed-fed tilapia (*Oreochromis niloticus*)

Abstract

The effects of prolonged exposure to sub-lethal un-ionised ammonia concentrations on the growth performance of juvenile Nile tilapia (*Oreochromis niloticus*) fed on fresh duckweed (*Lemna gibba*) grown on pre-treated domestic sewage has been investigated. The experiment was conducted in duplicates over 75 days using juveniles with a mean body weight of 20 g. Five nominal, total ammonia nitrogen concentrations (control, 2.5, 5, 7.5 and 10 mg N l^{-1}) were established as treatment groups. Statistical analysis of the specific growth rate (SGR) showed no significant ($p > 0.05$) differences between the SGR (0.71) of tilapia in the control (0.004 mg UIA-N l^{-1}) and the SGR (0.67) of those exposed to 0.068 mg UIA-N l^{-1}. The SGR of tilapia exposed to un-ionised ammonia nitrogen over 0.068 mg UIA-N l^{-1} (0.144, 0.262 and 0.434 mg UIA-N l^{-1}) was significantly reduced ($p< 0.01$). The non-observable effect concentration was 0.068 mg UIA-N l^{-1}, while the lowest observable effect concentration was 0.144 mg UIA-N l^{-1}. Increasing the un-ionised ammonia concentration increased the feed conversion ratio (FCR). At 0.144 mg UIA-N l^{-1}, the FCR increased by a factor of 1.6 of the value observed in the control, while at 0.262 mg UIA-N l^{-1} the FCR increased by a factor of 2.7. At 0.434 mg UIA-N l^{-1}, the FCR increased by a factor of 4.3. Protein efficiency ratio (PER) was also negatively correlated with the un-ionised ammonia concentration above 0.068 mgUIA-N l^{-1}. This study concluded that, for raising *O. niloticus* in fishponds fed on fresh duckweed, the toxic level of UIA-N and its negative effect on the growth performance, lies between 0.07 and 0.14 mg UIA-N l^{-1}. It is recommended that the un-ionised ammonia concentration be maintained below 0.1 mg UIA-N l^{-1}.

Key words: chronic ammonia toxicity, un-ionised ammonia, *Oreochromis niloticus*, duckweed, *Lemna gibba*, domestic sewage

1. Introduction

Ammonia is very toxic, not only to fish but all aquatic animals (Baird *et al.*, 1979; Zhao *et al.*, 1997; Harris *et al.*, 1998), especially in pond aquaculture at low concentrations of dissolved oxygen (Alabaster *et al.*, 1983). The toxic levels of un-ionised ammonia for short-term exposure usually are reported to lie between 0.6 and 2 mg/l, while some consider the maximum tolerable concentration to be 0.1 mg l^{-1} (Pillay, 1992). The concentration at which un-ionised ammonia is toxic becomes lower at decreasing dissolved oxygen concentrations in ponds (Alabaster *et al.*, 1983). Infectious dropsy of carp was an unknown-cause disease characterised by large quantities of mucous on the skin and gills and vascular congestion of the blood vessels in the scale-less skin and around the fins. Later on, a correlation between the disease and ammonia concentration in ponds was investigated and subsequently the disease was renamed environmental ammonoletic syndrome (Seymour, 1980). Feed efficiency and body composition of fish are negatively affected by ambient ammonia concentration. The main change in the body composition includes an increase in the water content in ammonia-exposed fish (Person-Le Ruyet *et al.*, 1997b).

Two primary processes affect ammonia concentration in ponds, the ammonia excretion rate by fish (indigenous ammonia) and sediment diffusion (Hargreaves, 1997) that represents one part of the exogenous ammonia in the pond. Ammonia concentration is sensitive to feeding rate, the protein content of feed, and its type and digestibility and the phytoplankton specific uptake rate. The production of indigenous ammonia is related to the quantity of nitrogen supplied by the protein contained in the feed (Buttle *et al.*, 1995; Brunty *et al.,* 1997, Chakraborty and Chakraborty, 1998; Thomas and Piedrahita, 1997). The protein catabolism pathway includes hydrolysis of long polypeptide chains into carboxylic acids via deamination and transamination enzymatic processes. In case of fish, there is no further transformation of ammonia and it is spontaneously lost in the water through the gill tissue. In tilapia, a significant increase in total ammonia nitrogen production was observed during the 24 hours period following feeding as the protein content of the feed increased (Brunty *et al.*, 1997). In *Clarias gariepinus*, small differences in relative protein may evoke distinct patterns of ammonia efflux over a range of dietary protein levels (Buttle *et al.*, 1995). Different species of fish and fish at different life stages within the same species are differentially able to utilise the dietary protein and the relationship between dietary protein levels and total ammonia production for fish species and life stages in the species may vary.

Ammonia toxicity is a problem relevant to the practice of fish farming, since the growth yield may be depressed by usual ambient ammonia concentrations, under intensive farming conditions (Person-Le Ruyet *et al.*, 1997a) and in sewage-fed aquaculture. It has been reported that high ammonia and low dissolved oxygen concentration during the summer and spring appeared to be the major factors responsible for mortality in sewage-fed fishponds (Wrigley *et al.*, 1988). Ammonia toxicity was the main reason of fish mortality in sewage oxidation ponds (El-Gohary *et al.*, 1995; Nasr *et al.*, 1998). In sewage-fed fish farms, sewage ammonia is the main, if not only, source of pond fertilisers. Algae utilise ammonia to grow and produce oxygen, whereas fish require both. The presence of more nitrogen loads to the pond increases both primary and secondary productivity of ponds and so more feed is available for fish for maximum growth (Diana *et al.*, 1997). On the contrary, algae increase biodegradable organic matter and consume more oxygen during the night and create water hypoxia (Pillay, 1992; Diana *et al.*, 1997). The algae bloom also elevates the water pH value and shifts the ammonia balance towards more un-ionised ammonia, which results in greater toxicity for fish (Pillay, 1992; Diana *et al.*, 1997). In sewage-fed fish aquaculture systems the ammonia concentration will increase, decrease, or remain stable over time if ammonia excretion by fish and ammonia in the inlet water, are greater than, less than, or equal to the nutrient assimilation by phytoplankton and nutrient losses, respectively (Seawright *et al.*, 1998). In the sewage-fed fish farm, a dynamic balance between ammonia and algae concentrations should be optimised for optimal growth conditions. A stocking density between 10,000 and 30,000 fingerlings per hectare in a sewage-fed pond is recommended with an optimal nitrogen loading rate of 4 Kg N/ha/day (Mara *et al.*, 1993).

World-wide use of duckweed in wastewater treatment means that it is available at a reasonable price for use in fish aquaculture since it has high protein content. However, ammonia causes moderate and severe physiological disturbance, depending on the level of ammonia and exposure duration. The chronic toxicity level of ammonia in duckweed-fed tilapia has to be defined since deficiency of duckweed in certain nutritional elements might affect the immune system and resistance of tilapia to un-ionised ammonia. The main objective of the current research was to study the effect of chronic ammonia toxicity on the growth performance of tilapia (*Oreochromis niloticus*) fed on fresh duckweed. The exposure-time was two and half months, which was more than the minimum of one month recommended for chronic toxicity tests on fish growth (APHA, 1998).

2. Material and methods

2.1. Fish species and test conditions

The experiment was conducted using juvenile tilapia (*Oreochromis niloticus*) with 20 g mean body weight, all from the same parental stock, obtained from a commercial fish farm, Cairo, Egypt. For two weeks prior to the experiment the fish were adapted to the pre-selected experimental conditions in one tank with an effective water volume of 480 l and 1m^2 surface area. The tank was supplied with running de-chlorinated and continuously aerated city tap water. The de-chlorination was performed using sodium thiosulfate solution (0.025M). The water flow rate was 50 l per day. Tilapia were fed with fresh duckweed harvested from a duckweed pond system treating effluent from an Up-flow Anaerobic Sludge Blanket (UASB) reactor fed with domestic sewage (El-Shafai *et al.*, submitted). The daily feeding rate was 250 gram fresh duckweed (*Lemna gibba*) per kg of fish. Daily maintenance and cleaning of the fish tank from faeces was done.

Two days before the experiment, the fish were randomly distributed between the experimental tanks. Each tank had 100 litre effective water volume and 0.26 m^2 surface area. The tanks were arranged in two sets of 5 tanks each. The maximum recommended fish densities for toxicity testing at a temperature > 20 ^{0}C is 2.5 g l^{-1} (APHA, 1998). In this experiment, initial stocking-density was four fish per each tank (20 g average weight each). The fish were allowed to adapt for two days, during which feeding was performed using a 25% feeding rate based on fresh duckweed (*Lemna gibba*) and live body weight of fish, Table 1. The tanks were fed with de-chlorinated city tap water at 25 litres daily.

2.2. Experimental design and analytical procedures

In addition to the control, four treatments with nominal total ammonia nitrogen (TAN) concentrations of 2.5, 5, 7.5 and 10 mg N l^{-1} were tested with duplicates for each treatment. Ammonium chloride stock solution (25 g l^{-1}) was used as a source of ammonia. Each tank was fed with de-chlorinated city tap water containing the desired ammonia concentrations. When necessary, the ammonia concentration in the water flow was adjusted to the desired concentration. The experiment lasted 75 days during which total feed intake was recorded for each treatment. The fish in each treatment were weighed monthly and at the end of the trial. Daily observation of the fish behaviour was performed.

TAN was determined using the boric acid-sulphuric acid titration method (APHA, 1998). Un-ionised ammonia nitrogen (UIA-N) concentrations were calculated using the general equation of bases (Albert, 1973):

$$NH_3 = \frac{[NH_3 + NH_4^+]}{[1 + 10^{(pKa-pH)}]} \qquad (1)$$

In fresh water, the calculation of pKa is based on the equation developed by Emerson *et al.* (1975):

$$PKa = 0.09018 + 2729.92/T \qquad (2)$$

(T = degree Kelvin = 273+ T°C)

Daily (at 10:30 am) pH was determined electrochemically using a "Radio pH meter model 62". BOD, nitrate nitrogen, nitrite nitrogen, total hardness and total alkalinity were measured according to the Standard Methods for the Examination of Water and Wastewater (APHA, 1998). BOD, nitrate nitrogen, total hardness and total alkalinity were measured weekly while nitrite was measured twice a week.

Table 1: Operational conditions in fish tanks

Item	1	2	3	4	5
Surface area of tank (m^2)	0.26	0.26	0.26	0.26	0.26
Working volume (l)	100	100	100	100	100
Initial mean body weight (g fish^{-1}) (average ± SD)	21.2 ± 5.5	21.5 ± 6.5	24.3 ± 7.4	26.2 ± 5.7	26.1± 9.3
Stocking density (fish tank^{-1})	4	4	4	4	4
Feeding rate (%)	25	25	25	25	25
Water flow (l d^{-1})	25	25	25	25	25

2.3. Studied parameters

Mortalities were counted daily for different exposure doses. Every month and at the end of the trail, the fish in each tank were weighed individually and specific growth rates (SGR) were calculated using the following expression:

$$SGR = \frac{(\ln Wf - \ln Wi)}{days} \times 100 \qquad (3)$$

Where, Wi and Wf are initial and final mean body weight respectively.

The total feed intake was recorded in each treatment, and the feed conversion ratio (FCR) and protein efficiency ratio (PER) were calculated based on the following expression:

$$FCR = \frac{\text{total feed ingested (dry weight)}}{\text{fish weight gain (wet weight)}} \qquad (4)$$

$$\text{Protein efficiency ratio (PER)} = \frac{\text{fish weight gain}}{\text{total protein ingested}} \qquad (5)$$

Composition of duckweed was analysed for dry matter and protein content. Dry matter content was determined by drying at 70^0C overnight. The protein content was calculated from the organic nitrogen (org-N × 6.25) using the macro-Kjeldhal method (APHA, 1998).

2.4. Data Analysis

All the results were presented as mean values and standard deviation. The influence of ammonia concentration was analysed by one-way analysis of variance (one way ANOVA). The lowest-observable-effect concentration (LOEC) was estimated (Person-Le Ruyet *et al.*, 1997a).

3. Results

Table 2: Water quality in fish tanks during chronic ammonia toxicity experiment

Parameter	1 (control)	2	3	4	5
Temperature °C	26-34	26-34	26-34	26-34	26-34
pH	7.2-7.5	7.3-7.6	7.3-7.7	7.4-7.8	7.4-7.9
BOD (mg O_2 l^{-1})	5.2±0.6	4.9±0.5	5.3±1.3	6.0±0.4	6.5±1.5
DO (mg O_2 l^{-1})	7.2±1	7.5±0.8	7.7±0.8	8.0±0.7	8.3±0.8
TAN[*] (mg N l^{-1})	0.2±0.04	2.9±0.7	5.3±0.6	7.5±0.5	10±0.6
UIA-N[**] (mg N l^{-1})	0.004±0.008	0.068±0.03	0.144±0.06	0.262+0.1	0.434±0.18
Nitrite (mg N l^{-1})	0.04±0.03	0.036±0.028	0.043±0.03	0.037±0.026	0.033±0.025
Nitrate (mg N l^{-1})	0.067±0.03	0.072±0.04	0.085±0.05	0.088±0.05	0.095±0.08
Total hardness (mg $CaCO_3$ l^{-1})	120±1	121±1	120±2	121±2	121±1
Total alkalinity	127±3	124±6	128±4	129±5	129±5

[*]Total Ammonia Nitrogen,
[**] Un-Ionised Ammonia Nitrogen

3.1 Water quality

The experiment extended for 75 days during which the temperature was the same in all tanks, but it varied during this period in the range of 26-34 °C. The pH values in the treatment tanks are presented in Table 2. The DO was more than 5 mgO_2 l^{-1} in all tanks. The calculated un-ionised ammonia based on the temperature, pH and total ammonia concentration was, 0.004, 0.068, 0.144, 0.262 and 0.434 mg UIA-N l^{-1} in the control (Treatment 1), treatment 2, 3, 4 and 5, respectively (Table 2). The total hardness and total alkalinity were stable. Except ammonia, no significant differences in the water quality were detected between the treatment tanks. Statistical analysis of water quality showed no significant differences between the parallel treatments in the two sets of experiment.

3.2. Growth performance

Growth performance of tilapia in different treatment tanks is presented in Table 3. No mortality was detected in any of the treatments during the experimental period. The results of the specific growth rates (SGR) showed mean values of 0.71, 0.67, 0.29, 0.2 and 0.14 in treatments 1, 2, 3, 4 and 5, respectively. The statistical analysis showed no significant difference (p>0.05) between the control group (0.004 mg UIA-N l^{-1}) and tilapia exposed to 0.068 mg UIA-N l^{-1}. The SGR of treatments 3 (0.144 mg UIA-N l^{-1}), 4 (0.262 mg UIA-N l^{-1}) and 5 (0.434 mg UIA-N l^{-1}) were significantly lower (p<0.01) than the control and treatment 2. There was a significant difference (p<0.05) between the SGR of tilapia exposed to 0.144 and 0.262 mg UIA-N l^{-1} but no difference (p>0.05) was detected between the SGR of tilapia in group 4 and those in group 5.

Table 3: Growth performance of tilapia exposed to chronic ammonia toxicity

Item	1(control)	2	3	4	5
Initial mean body weight (g fish^{-1})	21.2 ± 5.5	21.5 ± 6.5	24.3 ± 7.4	26.2 ± 5.7	26.1± 9.3
Final mean body weight (g fish^{-1})	36.2 ±15	35.5 ± 13.3	30.1 ± 9.5	30.4 ± 7.7	29 ± 10.8
Fish weight gain (g fish^{-1})	15	14	5.8	4.2	2.9
Total feed intake (g fish^{-1})**	22.3	22.1	22.6	23.6	23.2
SGR (average ± SD)	0.71±0.13^a	0.67±0.1^a	0.29±0.04^{b*}	0.2±0.05^{c*}	0.14±0.03^c
FCR (g feed g fish^{-1})	1.5	1.6	3.9	5.6	8
Total protein intake (g fish^{-1})	8.03	7.96	8.14	8.5	8.35
PER (g fish g protein^{-1})	1.78	1.76	0.71	0.49	0.35

P >0.05, p < 0.01

*p < 0.05

** g dry duckweed/fish

No effect on the feed intake was detected in any of the treatment groups. The feed conversion was affected by ammonia concentrations above 0.068 mg UIA-Nl^{-1}. There was no difference between the FCR of the control and group 2. The FCR increased by 4.3 fold in treatment 5 (0.434 mg UIA-Nl^{-1}) as compared to the control (Fig. 1). Increasing the UIA-N above 0.068 mg l^{-1} progressively decreased the protein efficiency ratio (Fig. 1). The negative effect of un-ionised ammonia on the growth performance of tilapia increased with the exposure time (Fig. 2).

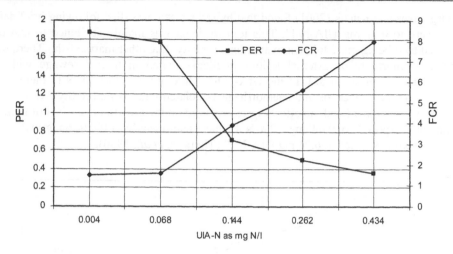

Fig. 1: PER and FCR in relation to UIA-N concentrations

4. Discussion

Of all environmental factors that control the fish production in fish farming system, dissolved oxygen concentration and ammonia concentration are the main factors. Low dissolved oxygen concentrations can be managed effectively by aeration. Ammonia is the main form of metabolic waste products of protein and so its practical management becomes more problematic, especially in ponds receiving high stocking densities, high feeding rates and/or fed with treated sewage. The un-ionised ammonia, which is the toxic form of ammonia, may limit fish production if the ammonia assimilation capacity of the pond is exceeded. Although, considerable information on the sensitivity of fish and aquatic animals to ammonia has been reported no available data for ammonia toxicity to tilapia fed on sewage grown duckweed.

In sewage oxidation ponds, fish mortality is mainly attributed to high un-ionised ammonia concentration and low dissolved oxygen concentrations (Wrigley *et al.*, 1988). El-Gohary *et al* (1995) studied fish production in sewage oxidation pond and reported 100% mortality for silver carp after 9 days, which was attributed to an un-ionised ammonia concentration of 0.41 mg N/l. In the current experiment no mortality was detected at exposure levels up to 0.434 mg UIA-N l[-1]. The low level of lethal ammonia toxicity reported by El-Gohary *et al.* (1995) was attributed to the high sensitivity of silver carp, since tilapia were stocked in the same pond with a calculated 0.52 mg UIA-N l[-1] for 92 days without any mortality. Long-term effects of constant exogenous ammonia concentrations were studied in two different batches of turbot juveniles (53 and 73 g) under controlled environmental and feeding condition (Person-Le Ruyet *et al.*, 1997a). Survival was maximal up to 0.33 mg UIA-N l[-1] while at 0.73 mg UIA-N l[-1] 50% mortality was observed on day 52 for 73 g turbot juveniles and on day 77 for 53 g turbot juveniles. They estimated a 28-day efficient concentration that gives 50% of the SGR of control between 0.6 and 0.75 mg UIA-N l[-1]. The chronic effects of ammonia on three batches of trout (14, 23, 104 g) juveniles indicated that there was no mortality up to 0.4 mg UIA-N l[-1] (Person-Le Ruyet *et al.*, 1997b).

Hargreaves and Kucuk (2001) found that hybrid striped bass died following brief daily exposure to 0.91 mg UIA-N l⁻¹. This toxicity level was mainly attributed to that the striped bass being more tolerant to ammonia toxicity, like other marine fish. There was no mass mortality or external lesions in tilapia reared in treated sewage with an ammonia concentration of 0.14 mg UIA-N 1-1 while at 0.45 mg UIA-N l⁻¹ there were clear skin ulcers, necrosis and haemorrhage associated with fish mortality (Nasr *et al.*, 1998). On the other hand, no effects were observed on the growth or survival of fathead minnows exposed to 0.44 mg UIA-N l⁻¹ but clear negative effects on growth and survival were detected at 0.91 mg UIA-N l⁻¹ (Thurston *et al.*, 1986).

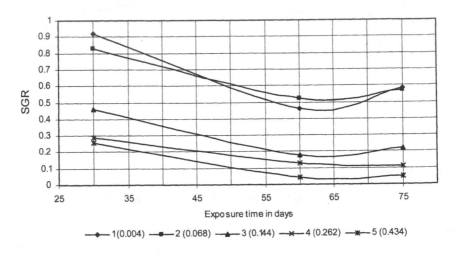

Fig. 2: Effect of exposure time on SGR of tilapia exposed to UIA-N

In our experiment, the un-ionised ammonia concentrations lower than 0.1 mgl⁻¹ did not affect the growth performance of tilapia. In fresh water crustaceans, *Eriocheir sinensis*, results (Zhao *et al.*, 1997) showed that the tolerance to ammonia increased as the larvae developed to juveniles and decreased by increasing the exposure time, which is similar to our results, Fig. 2. They reported that the 96h LC50 for juveniles is 0.9 mg UIA-N l-1 and they proposed a safe level of UIA-N for juveniles of 0.09 mg UIA-N l⁻¹. Our results showed that the lowest-observable effect concentration on the growth performance is 0.144 mg UIA-N 1-1. This value is similar to what has been reported (0.14 and 0.33 mg UIA-N l⁻¹) for turbot juveniles with a body weight 53 and 73 g (Person-Le Ruyet *et al.*, 1997a). It is also comparable to the range of 0.1-0.4 reported for turbot juveniles with a body weight range of 14-104 g (Person-Le Ruyet *et al.*, 1997b). Hargreaves and Kucuk (2001) reported that the growth of hybrid striped bass was not affected by brief daily exposure to 0.37 mg UIA-N l⁻¹. Brief daily exposure to 0.91 mg UIA-N l⁻¹ did not affect the growth performance of channel catfish or the blue tilapia. The toxicity effect of ammonia on the SGR is directly proportional with the exposure time (Fig. 2). An experiment that studied the growth performance of trout exposed to 0.025 and 0.033-mg UIA-N l⁻¹, showed no change in growth rate within 4 months, however, there was a significant reduction after 6 and 12 months at 0.033 mg UIA-N l⁻¹ (Smith and Piper, 1975). For juveniles of gilthead seabream (*Sparus aurata*), Wajsbrot *et al.* (1993) reported that the maximum acceptable toxic concentration for growth was between 0.27 and 0.47 mg UIA-N l⁻¹. The author concludes that *Sparus aurata* is less sensitive to ammonia than other marine and fresh water fish. Ammonia of 0.47 mg UIA-N l^{-1} and higher significantly suppressed the fish growth (p < 0.05) with clear liver function alteration. The growth rate of juvenile channel catfish was reduced during a 31 day growth trial when exposed to ammonia that ranged from 0.048 to 0.989 mg UIA-N l^{-1} (Colt and Tchobalongolous, 1978).

The non-effect observable concentration is the concentration of toxicant that can be present in water without affecting growth, reproduction or survival over the full life cycle of the test organism. The no-observable effect concentration in this study is 0.068 mg UIA-N l^{-1} that is lower than the recorded value (0.18-0.33) for turbot juveniles with a body weight range of 14-104 g, (Person-Le Ruyet *et al.*, 1997b). Szumski *et al.* (1982), suggested that a non-effect ammonia criterion of 0.08 mg UIA-N l^{-1} could be applied to warm water fish. During this experiment no reduction in the feed intake has been recorded in ammonia exposed fish up to 0.434 mg UIA-N l^{-1}. Person-Le Ruyet *et al* (1997b) reported that the first response to ammonia exposure was a change in turbot appetite, which was regulated within one week, but the full regulation of appetite, was limited to 0.4 mg UIA-N l^{-1}. Alderson (1979) reported that the threshold levels of un-ionised ammonia below which no effect on the growth performance could be observed was 0.066 mg UIA-N l^{-1} for Dover sole and 0.11 mg UIA-N l^{-1} for turbot.

The presence of ammonia was observed to decrease feed utilisation and increase the water content of the body and urea excretion rate in turbot juveniles (Person-Le Ruyet *et al.*, 1997a). Also Colt and Tchobalongolous, (1978) found an increase in the water content of channel catfish in an ammonia concentration range of 0.048-0.989 mg UIA-N l^{-1} with 50% reduction in growth at 0.517 mg UIA-N l^{-1}. Histopathological and haematological effects, particularly those affecting gills and liver function may contribute to reduce fish growth through inducing tissue hypoxia. Bucher and Hofer (1993) reported that the liver and kidney of brown trout reared in a mixture of ground water and treated sewage with an average un-ionised ammonia concentration of 0.13 mg UIA-N l^{-1}, showed tissue alteration. Clear histopathological alterations in gills, kidney and liver were detected in trout exposed to 0.067 and 0.076 mg UIA-N l^{-1} (Thurston *et al.*, 1984). Nasr *et al* (1998) reported that exposure to 0.33 mg UIA-N l^{-1} induced gill damage and subsequent liver tissue hypoxia associated with blood anaemia. Fivelstad *et al.* (1995) reported that the plasma glucose of Atlantic salmon was significantly increased in fish exposed to two ranges of un-ionised ammonia, medium (0.014-0.032 mg UIA-N l^{-1}) and high (0.043-0.08 UIA-N l^{-1}) and this increase was proposed to be due to enzymatic stimulation of glycolysis. The mean plasma glucose values were considered to be in the normal range for all groups.

Hargreaves and Kucuk (2001), proposed that the digestibility of dietary protein and the energy source might have been affected by un-ionised ammonia. The biochemistry of energy derivation from fats, carbohydrate and especially from protein is compromised by the presence of ammonia. Additionally, ammonia detoxification by fish is energy dependent and can result in a 68% reduction in the normal rate of energy production (Zieve, 1966). So, the effects of ammonia on energy utilisation may explain growth suppression in fish exposed to ammonia. Other mechanisms may contribute to growth reduction caused by ammonia exposure. Liver glycogen depletion and consequent blood acidosis (Sousa and Meade, 1977; Chetty *et al.*, 1980) may contribute to increased susceptibility to hypoxia.

5. Conclusions

The main conclusion of this study is the maintenance of an un-ionised ammonia concentration lower than 0.1 mg UIA-N l^{-1} in tilapia pond fed with fresh duckweed. There was no fish mortality at 0.144, 0.262 and 0.434 UIA-N l^{-1} while the growth rate was negatively affected at such levels.

Acknowledgements

The authors would like to thank the Dutch government for financial support for this research within the framework of the SAIL funded "Wasteval" project (LUW/MEA/971). The project is a co-operation between the Water Pollution Control Department (NRC, Egypt), Wageningen University and UNESCO-IHE Institute for Water Education, The Netherlands. The authors would like to express their sincere thanks to the technicians of the Water Pollution Control Department, NRC, Egypt.

References

Alabaster, J.S., Shurben, D.G., Malleit, M.J. 1983. The acute lethal toxicity of mixture of cyanide and ammonia to smolts of salmon, *Salmo salar* L. at low concentration of dissolved oxygen. J. Fish Biol. 22, 215-222

Albert, A. 1973. Selective toxicity. Chapman and Hall, London

Alderson, R. 1979. The effect of ammonia on the growth of juvenile Dover sole, *solea solea* (L.) and turbot, *Scophthalmus maximus* (L.). Aquaculture 17, 291-309

American Public Health Association 1998. Standard Methods for the Examination of Water and Wastewater, 20th edition, Washington.

Baird, R., Bottomley, J., Taitz, H. 1979. Ammonia toxicity and pH control in fish toxicity bioassays of treated wastewater. Water Res. 13, 181-184

Brunty, J.L., Bucklin, R.A., Davis, J., Baird, C.D., Noedstedt, R.A. 1997. The influence of feed protein intake on tilapia ammonia production. Aquacult. Eng. 16, 161-166

Bucher, F., Hofer, R. 1993. The effect of treated domestic sewage on three organs (gills, kidney, liver) of brown trout (*Salmo trutta*). Water Res. 27(2), 255-261

Buttle, L.G., Uglow, R.F., Cowx, I.G. 1995. Effect of dietary protein on the nitrogen excretion and growth of the African catfish, *Clarias gariepinus*. Aquat. Living Resour. 8, 407-414

Chakraborty, S.C., Chakraborty, S. 1998. Effect of dietary protein level on excretion of ammonia in Indian major carp, *Labeo rohita* fingerlings. Aquacult. Nutr. 4, 47-51.

Chetty, C.S., Naidu, R.C., Reddy, Y.S. 1980. Tolerance limits and detoxification mechanism in the fish, *Tilapia mosambica* to ammonia toxicity. Indian J. Fish. 27, 177-182

Colt, J., Tchobalongolous, G. 1978. Chronic exposure of channel catfish, *Ictalurus punctatus* to ammonia: effects on growth and survival. Aquaculture 15, 353-372.

Diana, J.S., Szyper, J.P., Balterson, T.R., Boyd, C.E. and Piedrahita, R.H. 1997. Water quality in ponds. In: Egna, H.S. and Boyd, C.E. (Eds.), Dynamics of pond aquaculture 1997, Lewis publishers in an imprint of CRC press, New York, USA, pp. 53-71.

Emerson, K.R., Russo, R.C., Lund, R.E., Thurston, R.V. 1975. Aquous ammonia equilibrium calculations: effect of pH and temperature. J. Fish. Res. Board Can. 32, 2377-2383

El-Gohary, F., El-Hawarry, S., Badr, S., Rashed, Y. 1995. Wastewater treatment and reuse for fish aquaculture. Water Sci. Technol. 32(11), 127-136.

El-Shafai, S.A., El-Gohary, F.A., Nasr, F.A., van der Steen, N.P. and Gijzen, H.J. (submitted) Nutrient recovery from domestic wastewater using a UASB-duckweed ponds system

Fivelstad, S., Schwarz, J., Stromsnes, H., Olsen, A.B. 1995. Sublethal effects and safe levels of ammonia in seawater for Atlantic salmon postsmolts (*Salmo salar* L.). Aquacult. Eng. 14(3), 271-280

Hargreaves, J.A. 1997. A simulation model of ammonia dynamics in commercial catfish ponds in the southeast United States. Aquacult. Eng. 16, 27-43

Hargreaves, J.A., Kucuk, S. 2001. Effects of diel un-ionised ammonia fluctuation on juvenile hybrid stripped bass, channel catfish and blue tilapia. Aquaculture 195, 163-181

Harris, J.O., Maguire, G.B., Edwards, S., Hindrum, S.M. 1998. Effect of ammonia on the growth rate and oxygen consumption of juvenile greenlip abalone, *Haliotis laevigata* Donovan. Aquaculture 160, 259-272

Mara D.D., Edwards P., Clark D. and Mills S.W. 1993. A rational approach to the design of wastewater-fed fishponds. Water Res. 27(12), 1797-1799.

Nasr, F.A., El-Shafai, S.A., Abo-Hegab, S. 1998. Suitability of treated domestic wastewater for raising *oreochromis niloticus*. Egypt. J. Zool. 31, 81-94

Person-Le Ruyet, J., Delbard, C., Chartois, H., Le Dellion, H. 1997a. Toxicity of ammonia to turbot juveniles: 1-effects on survival, growth and food utilisation. Aquat. Living Resour. 10, 307-314

Person-Le Ruyet, J., Galland, R., Roux, A., Chartois, H. 1997b. Chronic ammonia toxicity to juvenile turbot (*Scophthalmus maximus*). Aquaculture 154, 155-171

Pillay, T.V.R. 1992. Aquaculture and The Environment, First edition, 1992, pp. 189.

Seawright, D.E., Stickney, R.R., Walker, R.B. 1998. Nutrient dynamics in integrated aquaculture-hydroponics systems. Aquaculture 160, 215-237.

Seymour, E. A. 1980. The effects and control of algal blooms in fishponds. Aquaculture 19, 55-74.

Smith, G.R., Piper, R.G. 1975. Lesions associated with chronic exposure to ammonia. In: Ribelin, W.E., Migaki, G., (Eds.), Pathology of Fishes, The University of Wisconsin Press, Madison, Wisconsin, pp. 497-512.

Sousa, R.J., Meade, T.L. 1977. The influence of ammonia on the oxygen delivery system of Coho salmon haemoglobin. Comp. Biochem. Physiol. 58(A), 23-28

Szumski, D.S., Barton, D.A., Putnam, H.D., Polta, R.C. 1982. Evaluation of EPA un-ionised ammonia toxicity criteria. J. Water Poll. Control Federation 54, 282-291.

Thomas, S.L., Piedrahita, R.H. 1997. Oxygen consumption rates of white sturgeon under commercial culture conditions. Aquacult. Eng. 16, 227-237

Thurston, R.V., Russo, R.C., Lauedtke, R.T., Smith, C.E., Meyn, E. L., Chakoumakos, C., Wang, K. C., Brown, C.J.D. 1984. Chronic toxicity of ammonia to rainbow trout. Trans. Am. Fish. Soc. 113, 56-73

Thurston, R.V., Russo, R.C., Meyn, E.L., Zajdel, R.K., Smith, C.E. 1986. Chronic toxicity of ammonia to fathead minnows. Trans. Am. Fish. Soc. 115, 196-207

Wajsbrot, N., Gasith, A., Diamant, A., Popper, D.M. 1993. Chronic toxicity of ammonia to juvenile gilthead seabream *Sparus aurata* and related histopathological effects. J. Fish Biol. 42, 321-328

Wrigley, T.J., Toerien, D.F., Gaigher, I.G. 1988. Fish production in small oxidation ponds. Water Res. 22(10), 1279-1285

Zhao, J.H., Lam, T.J., Guo, J.Y. 1997. Acute toxicity of ammonia to the early stage larvae and juveniles of *Eriocheir sinensis* H. Milne-Edwards, 1853 (Decapoda: Grapsidae) reared in the laboratory. Aquacult. Res. 28, 517-525

Zieve, L. 1966. Pathogenesis of hepatic coma. Arch. Int. Med. 118, 211-223

Chapter Six

Microbial quality of tilapia reared in faecal contaminated ponds

Submitted to Environmental Research as:

Saber A. El-Shafai, Huub J.Gijzen, Fayza A. Nasr and Fatma A. El-Gohary. Microbial quality of tilapia reared in faecal contaminated ponds.

Microbial quality of tilapia reared in faecal contaminated ponds

Abstract

Microbial quality of tilapia reared in four faecal-contaminated fishponds was investigated. One of the fishponds (TDP) received treated sewage with an average faecal coliform count of 4×10^3 cfu/100ml, and was fed on fresh duckweed grown on treated sewage. Number of faecal coliform on duckweed biomass ranged between 4.1 $\times 10^2$ to 1.6 $\times 10^4$ cfu/g fresh weight. The second fishpond (TWP) received treated sewage and was fed on wheat bran. The third fishpond (FDP) received freshwater and was fed on the same duckweed. Pond 4 (SSP) received only settled sewage with an average faecal coliform count of 2.1 $\times 10^8$/100ml. The average count in the fishponds were 2.2×10^3, 1.7×10^3, 1.7×10^2 and 9.4×10^3 cfu/100ml in TDP, TWP, FDP and SSP, respectively. The FDP had significantly ($p < 0.05$) lower faecal coliform count than the treated sewage-fed ponds and SSP. Microbial quality of tilapia indicated that all tissue samples were contaminated with faecal coliform except the muscle tissues. Ranking of the faecal coliform contamination showed a decrease in the order intestine > gills > skin > liver. Poor water quality (ammonia and nitrite) in SSP resulted in statistically higher faecal coliform in fish organs of about one $\log_{10}$ than in treatments with good water quality. Pre-treatment of sewage is therefore recommended.

Key words: Microbial quality, *Oreochromis niloticus*, faecal coliform, sewage, duckweed, water quality

1. Introduction

Recycling of domestic sewage in fish farming and agriculture is an effective way of pollution control, which contributes to cost recovery and provides cheap protein food production. Reuse of treated sewage in fish farming has been applied in experimental system aquaria as well as in full-scale (Hejkal *et al.*, 1983; Polprasert *et al.*, 1984; Shereif *et al.*, 1995). Use of raw domestic sewage or effluent from treatment plants in fish farming is being applied in many Asian countries. The main objective of this practice is to fertilise the ponds and enhance natural, primary and secondary productivity. In 1989 at least two-thirds of the world production of farmed fish was coming from ponds fertilised with animal and human waste (Mara and Cairncross, 1989). Despite there are no proven cases of human bacterial disease being transmitted via fish culture using animal wastes or human sewage, in some countries the public health risks are the main reason for rejecting such reuse of waste. The expected potential health risks associated with wastewater recycling could be overcome if the wastes are treated effectively before reuse. Fish reared in treated domestic sewage has to be examined to ensure it is suitable for human consumption. In case of microbial contamination fish could be used as fishmeal for animal, fish and poultry nutrition. At low concentrations, micro-organisms are present on the fish surface, gills and general viscera and this might represent a source for cross-contamination during fish processing (Pillay, 1992). When present in low numbers, pathogens are not likely to penetrate into the fish muscles. At total bacterial counts greater than 5×10^4 cfu/ml of pond water, pathogens may be found in the muscles of fish (Mara and Cairncross, 1989). This threshold level represents the limit of the natural defence barriers of the fish. It is also reported that invasion of fish muscles by bacterial pathogens is likely to occur when fish is raised in ponds with faecal coliform and salmonellae concentration of $> 10^4$ and $> 10^5$ per 100 ml, respectively (Mara and Cairncross, 1989). The WHO (1989) reported that only limited experimental and field data on public/human health effects of sewage-fertilised aquaculture are available. As a tentative bacterial guideline, WHO recommended that faecal coliform levels in fishpond water should be below $10^3/100$ ml. In view of the dilution of wastewater that occurs in most ponds, the faecal coliform concentration can normally be maintained below the guideline value by treating wastewater to a level of 10^3-10^4 faecal coliform/100 ml. Because of the lack of evidence of a risk of bacterial and viral infections to farm workers and nearby residents, the WHO guidelines did not include a limit for faecal coliform in case of restricted irrigation (WHO, 1989). For restricted irrigation a guideline limit for exposure to faecal coliform was recommended to save farmers, their children and nearby residents from enteric viral and bacterial infections (Blumenthal *et al.*, 2000). The guideline limit has been set to 10^5 faecal coliform/100 ml in case of sprinkler irrigation and 10^3 faecal coliform/100 ml in case of flood irrigation. In case there are insufficient resources to meet this stricter guideline limit, a guideline limit of $\leq 10^5$ faecal coliform/100 ml should be supplemented by other health protection measures. The restricted irrigation guideline was set to crops eaten after cooking. A similar guideline may be applicable to fish eaten cooked. The WHO guideline for reuse of treated sewage in fish aquaculture was established based on available knowledge and further research on the microbiology and epidemiology of wastewater fed aquaculture was recommended (WHO, 1989).

The main objective of this study is to evaluate and compare the microbial quality of tilapia (*Oreochromis niloticus*) stocked in different sewage-contaminated ponds. The effect of water quality (ammonia and nitrite) on fish contamination was discussed. An assessment was made of the effect of using treated sewage and harvested duckweed from a UASB-duckweed ponds system treating domestic sewage on the microbiological quality of tilapia.

2. Materials and methods

2.1. Fish species and husbandry

The experiment was carried out in the wastewater treatment research station of the Water Pollution Control Department, National Research Centre in Egypt. Nile tilapia (*Oreochromis niloticus*), weighing about 20 g, was used in the experiment. Tilapia juveniles were obtained from a hatchery fish farm, having the same parental stock. Two weeks before the experiment the fish were adapted to the selected experimental conditions in two plastic holding tanks. Each tank had a total surface area of 1m^2 and an effective water depth of 48 cm. The tanks were fed with de-chlorinated city tap water. Removal of chlorine was achieved by using sodium thiosulfate solution (0.025M) and overnight aeration using a compressor. Tilapia was stocked at a density of 50 fish per tank under continuous aeration and regular removal of faeces and uneaten feed. The water flow was kept at 50 litres per day. The fish were fed on a mixture of wheat bran and fresh duckweed to satiation.

2.2. Water sources and fish feed

The effect of water quality and fish feed was tested in four fishponds. Three water sources with different qualities were used. De-chlorinated city tap water was used as a freshwater source. The freshwater source was free of any pollutants and met drinking water standards. 2-hours settled sewage and effluent of a UASB-duckweed ponds (El-Shafai *et al.*, submitted) treating domestic sewage were used as sources of domestic sewage with different qualities. The effluent of the UASB-duckweed ponds met the standard for restricted irrigation. Water quality parameters of both settled and treated sewage were presented in Table 1. Two sources of fish feed were used, local wheat bran and fresh duckweed grown in pilot-scale sewage stabilisation ponds. The duckweed was harvested daily from three duckweed ponds in series. Harvesting from the same pond was done once every three days. Fresh duckweed was used as fish feed without further processing.

Table 1. Water quality parameters in the treated effluent and settled sewage (average ± standard deviation)

Parameters	Unit	Treated sewage	Settled sewage
Temperature	°C	24-34	25-31
pH	-	7.2-8.3	6.6-7.5
BOD	mg O_2 / l	14±5	208±46
TAN[1]	mg N / l	0.4±0.9	17.9±3.3
Nitrite nitrogen	mg N / l	0.17±0.1	0.01±0.02
Nitrate nitrogen	mg N / l	0.70±0.68	0.13± 0.03
TKN	mg N / l	4.4±1.4	23.2±5.4
Total phosphorus	mg P / l	1.11±0.41	3.58±0.84
TSS	mg / l	32±10	161±74
Dissolved oxygen	mg O_2 / l	6.3±1.2	Not detected
Faecal coliform	cfu / 100ml	$4 \times 10^3 \pm 3.7 \times 10^3$	$2.1 \times 10^8 \pm 1.5 \times 10^8$

[1] (TAN) total ammonia nitrogen

2.3. Experimental design and operating conditions

The experiment consisted of 4 ponds, which were operated for 150 days, from early June to early November with ambient temperature range of 20-38 °C. Each pond occupies 1 m^2 surface area and 48 cm water depth. At the start of the experiment all fishponds were filled with the respective water sources. The type of water source, flow rate, feed source and operating conditions in fishponds are presented in Table 2. In SSP, sewage was settled for 2 hours before transfer to the pond and flow rate was based on the nitrogen content of the settled sewage to provide a recommended N loading rate of about 4 kgN/ha/d (Mara *et al.*, 1993). The TWP was fed on wheat bran at a dry-weight feeding rate similar to the duckweed-fed ponds. The SSP received only settled sewage. Water quality parameters of the treated and settled sewage were presented in Table 1 as mentioned before.

Table 2: Operating conditions in fishponds

Item	TDP[2]	TWP[2]	FDP[2]	SSP[2]
Fish weight (g fish^{-1})	21.41±2.41	20.15±3.57	19.8±3.98	20.35±2.69
Stocking density (fish tank^{-1})	10	10	10	10
Feed source	Duckweed	Wheat bran	Duckweed	-
Daily feeding rate[1]	25	1.34-1.52	25	-
Water source	Treated sewage	Treated sewage	Tap water	Settled sewage
Flow rate (l d^{-1})	43	43	38	15

[1] gram feed per 100 gram fish

[2] TDP is treated sewage-duckweed-fed pond, TWP is treated sewage-wheat bran-fed pond, FDP is freshwater-duckweed-fed pond and SSP is settled sewage-fed pond.

2.4. Studied parameters and analytical procedures

During the experimental period samples were taken from incoming water and fishponds water for analysis of physical, chemical and microbiological parameters. The physical and chemical measurements included water temperature, pH, dissolved oxygen concentration (DO), biochemical oxygen demand (BOD), total ammonia nitrogen ($NH_3 + NH_4^+$), un-ionised ammonia nitrogen (NH_3), nitrite nitrogen, nitrate nitrogen, TKN, total phosphorous and TSS. All analyses were performed according to standard methods (APHA 1998).

The un-ionised ammonia nitrogen (UIA-N) concentrations were calculated using the general equation of bases (Albert, 1973):

$$NH_3 = \frac{[NH_3 + NH_4^+]}{[1 + 10^{(pKa-pH)}]} \tag{1}$$

In fresh water the calculation of pKa is based on the equation developed by Emerson *et al.* (1975).

$$pKa = 0.09018 + 2729.92/T \tag{2}$$

T = degree Kelvin

2.5. Microbiological study

The treated effluent, settled sewage and the water in the fishponds were subjected to microbiological investigation using faecal coliform as faecal pollution indicator. Water samples were collected in sterile test tubes, covered and transferred to the laboratory within minutes. The faecal coliform were counted by membrane filter technique and poured plate technique using membrane faecal coliform (m Fc) medium (APHA, 1998). Microbial quality of the duckweed feed was assessed by taking 2 grams of fresh duckweed, grinded in a sterile vortex, blended and serially diluted with sodium chloride saline solution (0.09%). Faecal coliform were counted using m Fc medium and poured plate technique. Microbial quality of the fish was investigated by taking samples of the skin, gills, intestine, liver and muscles. Three fishes were taken from each fishpond and killed by a blow on the head. A swab was taken from a 6 cm^2 surface area including the mucous and scales. The swab was diluted in 100 ml sterile sodium chloride solution (0.09%). The muscles sample was taken by sterilising the skin with 96% ethanol. Five grams of muscle tissue was taken and grinded in a sterile vortex, and then diluted to 100-ml using sodium chloride solution. The gills were completely removed, weighed and grinded in a sterile vortex followed by dilution to 100 ml using the saline solution. Also the intestine and liver were removed completely, weighed, grinded and diluted to 100 ml. All samples were vigorously shaken and the supernatant was taken and serially diluted, except for the muscle sample, which was analysed without dilution.

2.6. Data Analysis

All the results were presented as mean values and standard deviation. Statistical analysis between the treatments was done using one-way analysis of variance (one way ANOVA).

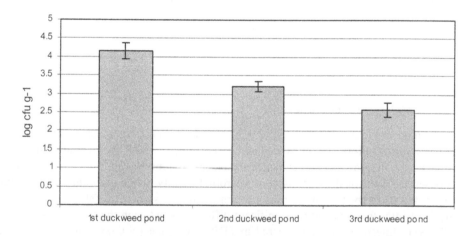

Fig 1: Number of faecal coliform per g of fresh duckweed biomass derived from 3 duckweed sewage treatment ponds placed in series

3. Results

3.1. Water quality and duckweed biomass

As presented in Table 1, the mean faecal coliform count in the incoming treated sewage and settled sewage was 4×10^3 cfu/100 ml and 2.1×10^8 cfu/100 ml, respectively. Microbial quality of fresh duckweed is presented in Figure 1. Table 3 shows some water quality parameters in the fishponds. The quality of water in fishponds shows higher concentration of total ammonia ($p<0.05$), un-ionised ammonia ($p<0.05$) and nitrite ($p<0.01$) in the settled sewage-fed pond (SSP) compared to other ponds. Faecal coliform concentrations in fishponds were 2.2×10^3, 1.7×10^3, 1.7×10^2 and $9.4.\times10^3$ in TDP, TWP, FDP and SSP, respectively. Except for FDP, results showed no significant ($p > 0.05$) differences between the microbial qualities of water in the ponds. The FDP contained significantly lower values than the treated sewage-fed ponds and SSP ($p < 0.05$).

Table 3: Water quality parameters in fishponds

Parameter	TDP	TWP	FDP	SSP
Temperature °C	23-35	23-35	23-35	23-35
pH	8-9.4	8-9.7	8.4-9.9	8.1-10.7
[1]Total ammonia (mg N / l)	0.22 ± 0.29^a	0.15 ± 0.2^a	0.12 ± 0.2^a	0.57 ± 0.89^c
[1]UIA-N (mg N / l)	0.06 ± 0.12^a	0.04 ± 0.08^a	0.05 ± 0.11^a	0.12 ± 0.22^c
[2]Nitrite nitrogen (mg N / l)	0.024 ± 0.012^a	0.018 ± 0.016^a	0.018 ± 0.011^a	0.112 ± 0.150^c
Nitrate nitrogen (mg N / l)	0.18 ± 0.07	0.13 ± 0.06	0.10 ± 0.06	0.14 ± 0.06
Dissolved oxygen (mg O_2/ l)[*]	15.9 ± 4.7	14.9 ± 4.7	15.3 ± 4.1	27.3 ± 8.1
[1]Faecal coliform (cfu / 100ml)	$(2.2\times10^3 \pm 1.6\times10^3)^a$	$(1.7\times10^3 \pm 1.5\times10^3)^a$	$(1.7\times10^2 \pm 2\times10^2)^b$	$(9.4.\times10^3 \pm 1.1\times10^4)^a$

[1] $P < 0.05$,

[2] $P < 0.01$ Values with the same superscript letters are statistically not significantly different (a, b and c),

[*] DO was not controlled and measured daily at 11:30 a. m.

3.2. Microbial quality of the fish

As presented in Fig 2, results of faecal coliform count in different fish organs showed that the faecal coliform count is higher by one $\log_{10}$ in SSP as compared to other treatments. In all treatments no evidence for faecal contamination in the muscle tissues was observed. Statistical comparison of microbial quality of the individual fish organs between the different treatments was performed and is shown in Table 4. Statistical analysis of intestines showed no significant differences between the treatments, except for the SSP. Intestines of tilapia reared in TDP had higher ($p<0.05$) number of faecal coliform than those reared in TWP and FDP. In case of gills, there was a significantly lower value in tilapia reared in FDP than those reared in TDP but there were no differences neither between TDP and TWP ($p > 0.05$) nor between FDP and TWP ($p > 0.05$).

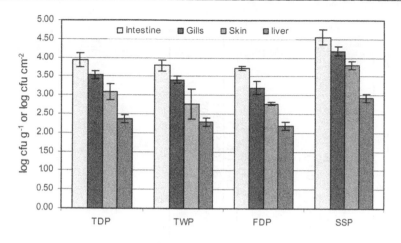

Fig. 2: Faecal coliform count in fish organs (log cfu/g or log cfu/cm²)

Generally, faecal coliform count in fish organs declined in the following order: intestine > gills > skin > liver. Statistical comparison of faecal coliform count in different fish organs within individual fishponds is presented in Table 5. In TDP no significant difference ($p > 0.05$) was observed neither between intestine and gills nor between skin and liver. In case of treated sewage-wheat bran-fed pond (TWP) significantly higher ($p < 0.05$) values of faecal coliform count were detected in the intestine as compared to the gills, but no significant difference was present between the skin and the liver ($p > 0.05$). In this pond a significantly higher faecal coliform count was found in intestine and gills as compared to skin and liver. In case of FDP there was a significantly higher count of faecal coliform in the intestine than in the other organs. In case of SSP the results showed no significant difference ($p > 0.05$) between the intestine and the gills, however, significantly higher values of faecal coliform were present in the previous organs as compared to the skin and the liver.

4. Discussion

Despite the absence of a significant difference between the faecal coliform counts in the water of SSP and other treatments, higher contamination of fish organs was demonstrated in that pond. This result might be attributed mainly to the significantly higher concentrations of total ammonia, un-ionised ammonia (UIA-N) and nitrite in the SSP, (Table 3). The poor water quality may have induced weakness of the fish resulting in higher susceptibility to bacterial infection (Escher *et al.*, 1999). The poor quality of water induces stress that is manifested in elevated cortisone levels, a hormone known to be a very potent immuno-suppressant (Pickering and Pottinger, 1989). The exposure of fish to un-ionised ammonia induces reduction in the spleenic lymphoid tissue (Smith and Piper, 1975) and may result in immune-deficiency. The stress effect of low dissolved oxygen concentration and high nitrite concentrations could also have resulted in increased susceptibility of fish to bacterial infection (Bunch and Bejerano, 1997)

Table 4: Statistical comparison[1] of faecal coliform count, in individual fish organs between treatment ponds

Fish organ	TDP	TWP	FDP	SSP
Intestine (cfu.g^{-1})	$(9.3\times10^3\pm4.2\times10^3)^{ab}$	$(6.5\times10^3\pm2.3\times10^3)^a$	$(5.3\times10^3\pm7.6\times10^2)^a$	$(3.8\times10^4\pm1.9\times10^4)^b$
Gills (cfu.g^{-1})	$(3.5\times10^3\pm8.6\times10^2)^a$	$(2.6\times10^3\pm5.3\times10^2)^{ab}$	$(1.7\times10^3\pm5.9\times10^2)^b$	$(1.6\times10^4\pm4\times10^3)^c$
Skin (cfu.cm^{-2})	$(1.3\times10^3\pm7\times10^2)^a$	$(7.2\times10^2\pm4.5\times10^2)^a$	$(6\times10^2\pm5\times10)^a$	$(6.5\times10^3\pm1.6\times10^3)^b$
Liver (cfu.g^{-1})	$(2.3\times10^2\pm5.9\times10)^a$	$(1.9\times10^2\pm4.9\times10)^a$	$(1.6\times10^2\pm4\times10)^a$	$(8.6\times10^2\pm2.1\times10^2)^b$

[1] Comparison is between values in the same row and values with the same superscripts are statistically not significantly different (P > 0.05)

Table 5: Statistical comparison[1] of faecal coliform count, in fish organs within individual fishponds

Treatment pond	Intestine (cfu.g^{-1})	Gills (cfu.g^{-1})	Skin (cfu.cm^{-2})	Liver (cfu.g^{-1})
TDP	$(9.3\times10^3\pm4.2\times10^3)^a$	$(3.5\times10^3\pm8.6\times10^2)^a$	$(1.3\times10^3\pm7\times10^2)^b$	$(2.3\times10^2\pm5.9\times10)^b$
TWP	$(6.5\times10^3\pm2.3\times10^3)^a$	$(2.6\times10^3\pm5.3\times10^2)^b$	$(7.2\times10^2\pm4.5\times10^2)^c$	$(1.9\times10^2\pm4.9\times10)^c$
FDP	$(5.3\times10^3\pm7.6\times10^2)^a$	$(1.7\times10^3\pm5.9\times10^2)^b$	$(6\times10^2\pm5\times10)^c$	$(1.6\times10^2\pm4\times10)^d$
SSP	$(3.8\times10^4\pm1.9\times10^4)^a$	$(1.6\times10^4\pm4\times10^3)^b$	$(6.5\times10^3\pm1.6\times10^3)^b$	$(8.6\times10^2\pm2.1\times10^2)^c$

[1] Comparison is between values in the same row and values with the same letter are statistically not significantly different (P > 0.05)

103

In this study higher numbers of faecal coliform were found in the intestines than in other organs in TWP and FDP while in TDP and SSP the significantly higher contamination was detected in both intestine and gills as compared to the skin and liver. The higher count in the gut could be attributed to the fact that the gut functions like a bioreactor and faecal coliform form part of the natural gut flora (APHA, 1998). The higher contamination of gills in comparison to the skin may be attributed to the structure of the gills, which has high specific surface area for bacterial attachment, and to the high water flow rate passing through it. Fatal et al. (1993) reported that tilapia reared in animal and human waste contaminated ponds, showed higher contamination in the digestive tract than the skin and liver while very few colonies were detected in the muscle. Similarly Hejkal et al. (1983) found higher contamination in the gut followed by the skin. On the other hand, by using primary treated sewage Khalil and Hussein (1997) reported higher contamination of gills of tilapia than intestine and skin. Ogbondeminu and Okoye, (1992) reported that in raw sewage fertilised ponds, faecal coliform numbers in tilapia decreased in the order skin > gills > intestine > muscle. In this experiment Faecal coliform count in the intestine and gills was higher than the concentration in the surrounding water. Fatal et al. (1993) reported higher numbers of faecal indicators and enterococci in the digestive tract of tilapia than in the surrounding water. In septage-fed fishponds, the results proved that all counts of faecal coliform in tilapia were of the same order of magnitude as counts in the water in which fish were cultured (Eves et al., 1995).

Naturally, there is a wide range of bacteria present on the skin of fish, which reflects the microbial composition of pond water. Several authors demonstrated a correlation between fish biomass and faecal coliform concentration in the pond (Markosova and Jezek, 1994; Davis et al., 1995). Also fish living in the natural environment are known to harbour enterobacteriacea that may cause diseases for humans and other warm-blooded animals (Pillay, 1992). Although in the natural environment the fish usually harbour bacteria only in the digestive tract (Polprasert et al., 1984) Vibrio species have been isolated and described from normal healthy Penaeus vannamei juveniles (Gomez-Gil et al., 1998). As well bacteria have been isolated from 14.3% of the animals with a count in the range of 2×10^2 - 3×10^3 cfu/ml of hymolymph. Streptococcus sp. was isolated from healthy and diseased tilapia, healthy common carp and diseased mullet and striped hybrid bass (Bunch and Bejerano, 1997). Isolation of aeromonas and streptococcus from a healthy population of rainbow trout was also demonstrated (Barham et al., 1979). Easa et al. (1995) found that faecal coliform on the skin of natural fish were higher than the count on tilapia stocked in sewage-fed ponds. At low numbers, microorganisms will be present on the surface of the fish and gut but not in the muscle tissue. Above a certain threshold level, which represents the limit of the natural defence mechanisms of the fish, pathogens are capable to penetrate into the muscle. Polprasert et al. (1984) concluded that the Escherichia coli threshold concentration beyond which pathogens could penetrate into the muscle is 10^4 cfu /100ml.

No faecal coliform contamination was detected in the muscle of tilapia even in the SSP at faecal coliform concentration of 9.4×10^3 cfu/100ml. Fatal et al. (1993) demonstrated E. coli at 0.2 log/g in muscles of tilapia reared in sewage-fed pond with E coli count of 10^5 cfu/100ml. Pillay (1992) reported that the faecal streptococci and other human disease-related bacteria were found in the gut content of pacific salmon and rainbow trout grown in domestic sewage-fertilised pond. No muscle contamination was detected in tilapia stocked in primary treated sewage with faecal coliform count of 5×10^3 (Khalil and Hussein, 1997). On the other hand microbial quality of tilapia and common carp stocked in untreated sewage-fed pond showed contamination of muscle by faecal coliform and faecal streptococci even when the count in the water less than 10^4 cfu/100ml (Ogbondeminu and Okoye, 1992). The public health officers and microbiologist within the United Kingdom and Ireland set guidelines for microbiological quality of some ready-to-eat foods. The guidelines for fish eaten raw or cooked are shown in Table 6. The test for enterobacteriaceae has replaced the test for faecal coliform that traditionally was used as indicator of hygiene and contamination. Except for the gut and gills of tilapia reared in the SSP the data meet the guideline. Safety precautions during fish processing are needed to avoid cross contamination. There seems to be no public health risk from fish reared in treated sewage from UASB-duckweed ponds or fed on fresh duckweed grown on UASB effluent. Similar results were reported by several demonstration projects (Easa *et al.*, 1995; Eves *et al.*, 1995; Pillay, 1992). Fatal *et al.* (1993) recommended that fish has to be cooked well before human consumption and the major public health concern could be the risk of *Aeromonas* wound infection among the workers who handle and process the fish. For further reduction of public health risks, purification of sewage-raised fish prior to sale was recommended. Richards (1988) reported that the potential advantages of the purification process are the removal of pathogens, objectionable odours and chemical contaminants.

Table 6: Guidelines for microbiological quality of some ready-to-eat foods (Gilbert *et al.*, 2000)

Criterion	Microbiological quality (cfu per gram)[*]			
	Satisfactory[3]	Acceptable[3]	Unsatisfactory[3]	Unacceptable[3]
Aerobic colony count 30°C/48h				
Fish eaten raw	$< 10^3$	$10^3 < 10^4$	$\geq 10^4$	Not applicable
Fish eaten cooked	$< 10^5$	$10^5 < 10^6$	$\geq 10^6$	Not applicable
Indicator organisms (for raw and cooked fish)				
Enterobacteriaceae	< 100	$100 < 10^4$	$\geq 10^4$	Not applicable
E. coli (total)	< 20	$20 < 100$	≥ 100	Not applicable
Listeria spp (total)	< 20	$20 < 100$	≥ 100	Not applicable
Pathogens (for raw and cooked fish)				
Salmonella spp	ND[1]	ND[1]	ND[1]	Detected
Campylobacter spp	ND[1]	ND[1]	ND[1]	Detected
E. coli O157 & VTEC[2]	ND[1]	ND[1]	ND[1]	Detected
Vibrio cholera	ND[1]	ND[1]	ND[1]	Detected

[1] not detected in 25 grams sample

[2] verocytotoxin producing *E. coli*

[3] Satisfactory means good microbiological quality, acceptable means a borderline limit of microbiological quality, unsatisfactory means further sampling is necessary, Unacceptable means potential hazards

[*] colony forming unit per gram

5. Conclusions

1· Tilapia reared in fishponds fed with water and/or duckweed coming from a UASB-duckweed ponds system treating domestic sewage is safe for human consumption and the only concern might come from cross-contamination from highly contaminated organs (general viscera, gills and skin) during fish processing.

2· Water quality parameters such as ammonia and nitrite in fishponds may affect the level of bacterial contamination of fish organs, probably due to the effect of water quality on the fish immune response.

3· The lower water quality in settled sewage fed ponds increased susceptibility of fish to bacterial infection and it is therefore recommended to pre-treat sewage before using it in fish aquaculture.

Acknowledgements

The authors would like to thank the Dutch government for financial support for this research, which was developed within the framework of the "Wasteval" project. The project is a co-operation between the Water Pollution Control Department (NRC, Egypt), Wageningen University, and UNESCO-IHE Institute for Water Education, The Netherlands. The authors thank Dr Henk Lubberding and Dr Peter van der Steen, Environmental Resource Department, UNESCO-IHE, for their help during this work. The authors would like to express his sincere thanks to the technicians of Water Pollution Control Department, NRC, Egypt for analytical support.

References

Albert, A., 1973. Selective toxicity. Chapman and Hall, London.

American Public Health Association, 1998. Standard Methods for the Examination of Water and Wastewater, 20th edition, Washington.

Barham, W.T., Schoonbee, H. and Smit, G.L., 1979. The occurrence of Aeromonas and streptococcus in Rainbow trout, *Salmo gairdneri* Richardson. J. Fish Biol. 15, 457-460.

Blumenthal, U.J., Mara, D.D., Peasey, A., Ruiz-Palacios, G. and Stoot, R., 2000. Guidelines for the microbiological quality of treated wastewater used in agriculture: recommendations for revising WHO guidelines. Bull. World Health Organ. 78(9), 1104-1116.

Bunch, E.C. and Bejerano, I., 1997. The effect of environmental factors on the susceptibility of hybrid tilapia *Oreochromis niloticus* X *Oreochromis aureus* to streptoccosis. Bamidgeh, 49(2), 67-76.

Davis, E.M., Mathewson, J.J. and De La Cruz, A.T., 1995. Growth of indicator bacteria in a flow-through aquaculture facility. Water Res. 29(2), 2591-2593.

Easa, M.E., Shereif, M.M., Shaaban, A.I. and Mancy, K.H., 1995. Public health implication of wastewater reuse for fish production. Water Sci. Technol. 32 (11), 145-152.

Emerson, K.R., Russo, R.C., Lund, R.E. and Thurston, R.V., 1975. Aquous ammonia equilibrium calculations: effect of pH and temperature. J. Fish. Res. Board of Canada 32, 2377-2383.

Escher, M., Wahli, T., Buttner, S., Meicr, W. and Burkhardt-Hom, 1999. The effect of sewage plant effluent on brown trout (*Salmo trutta* Fabrio): a cage experiment. Aquat. Sci. 61, 93-110.

Eves, A., Turner, C., Yakupitiyage, A., Tongolle, N. and Ponza, S., 1995. The microbiological and sensory quality of Nile tilapia (*Oreochromis niloticus*). Aquaculture 132, 261-272.

El-Shafai, S.A., El-Gohary, F.A., Nasr, F.A., van der Steen, N.P. and Gijzen, H.J. (submitted). Nutrient recovery from domestic wastewater using a UASB-duckweed ponds system

Fattal, B., Dotan, A., Parpari, L., Tchorsh, Y. and Cabelli, V.J., 1993. Microbiological purification of fish grown in faecally contaminated commercial fishpond. Water Sci. Technol. 27(7-8), 303-311.

Gilbert, R.J., de Louvois, J., Donovan, T., Little, C., Nye, K., Ribeiro, CD., Richards, J., Roberts, D. and Bolton, F.J., 2000. Guidelines for the microbiological quality of some ready-to-eat foods sampled at the point of sale. Commun. Dis. Public Health 3(3), 163-167.

Gomez-Gil, B., Tron-Mayen, L., Roque, A., Turnbull, J.F., Inglis, V. and Guerra –Flore, A.L., 1998. Species of vibrio isolated from hepatopancrease, hymolymph and digestive tract of a population of health juvenile *Panaeus vannamei*. Aquaculture 163, 1-9.

Hejkal, T.W., Gerba, C.P., Henderson, S. and Freeze, M., 1983. Bacteriological, Virological and Chemical evaluation of wastewater-aquaculture system. Water Res. 17(12), 1749-1755.

Khalil, M.T. and Hussein, H.A., 1997. Use of wastewater for aquaculture: an experimental field study at a sewage treatment plant, Egypt. Aquacult. Res. 28, 859-865.

Mara, D. and Cairncross, S., 1989. Guidelines for the safe use of wastewater and excreta in agriculture and aquaculture: measures for public health protection. World Health Organization; Geneva. 1989.

Mara, D.D., Edwards, P., Clark, D. and Mills, S.W., 1993. A rational approach to the design of wastewater-fed fishponds. Water Res. 27(12), 1797-1799.

Markosova, R. and Jezek, J., 1994. Indicator bacteria and limnological parameters in fish ponds. Water Res. 28(12), 2477-2485.

Ogbondeminu, F.S. and Okoye, F.C., 1992. Microbiological evaluation of an untreated domestic wastewater aquaculture system. J. Aquacult. Trop. 7(1), 27-34.

Pickering, A.D. and Pottinger, T.G., 1989. Stress responses and disease resistance in salmonid fish: Effects of chronic elevation of plasma cortisol. Fish Physiol. Biochem. 7, 253-258.

Pillay, T.V.R., 1992. Water and wastewater use. In "Aquaculture and The Environment" (T.V.R. Pillay, Ed.), pp. 49-55.

Polprasert, C., Udom, S. and Choudry, K.H., 1984. Septage disposal in waste recycling ponds. Water Res. 18(5), 519-528.

Richards, G.P., 1988. Microbial purification of shellfish: a review of depuration and relaying. J. Food Prot. 51, 218-251.

Shereif, M.M., Easa, M.E., El-Samra, M.I. and Mancy, K.H., 1995. A demonstration of wastewater treatment for reuse applications in fish production and irrigation in Suez, Egypt. Water Sci. Technol. 32 (11), 137-144.

Smith, G.R. and Piper, R.G., 1975. Lesions associated with chronic exposure to ammonia. In "Pathology of Fishes" (W.E. Ribelin and G. Migaki, Ed.), pp. 497-512. The university of Wisconsin press, Madison, Wisconsin

WHO, 1989. Health guidelines for the use of wastewater in agriculture and aquaculture. Technical Report Series, No. 778, World Health Organisation, Geneva.

Chapter Seven

Reuse of treated domestic sewage and duckweed (*Lemna gibba*) in tilapia aquaculture

Submitted to Bioresource Technology as:

Saber A. El-Shafai, Fatma A. El-Gohary, Johan A.J. Verreth and Huub J. Gijzen. Reuse of treated domestic sewage and duckweed (*Lemna gibba*) in tilapia aquaculture

Reuse of treated domestic sewage and duckweed (*Lemna gibba*) in tilapia aquaculture

Abstract

Reuse of treated domestic sewage and fresh duckweed (*Lemna gibba*) harvested from duckweed sewage stabilisation ponds has been applied in culture of Nile tilapia (*Oreochromis niloticus*). Fresh duckweed was compared with local fishmeal-based diet, while using treated sewage and de-chlorinated city tap water as a water source. The daily feeding rate in duckweed-fed fishponds was 25 g fresh duckweed per 100 g of live fish weight. The fishponds fed commercial feed received equal amounts of protein as in the duckweed-fed ponds. The growth performance of juvenile tilapia showed no significant ($p > 0.05$) difference in the Specific Growth Rate (SGR) of tilapia reared in treated sewage and fed either fresh duckweed or commercial feed. In freshwater ponds, the juveniles of tilapia fed on commercial feed had significant ($p < 0.01$) higher SGR than those fed on fresh duckweed. In case of duckweed fed ponds, the SGR of tilapia reared in treated effluent was significantly ($p < 0.01$) higher than in freshwater ponds. Similar results were obtained with commercial feed. The SGRs of tilapia were 0.97 and 1.01 in the treated sewage-duckweed-fed pond (TDP) and treated sewage-commercial feed-fed pond (TCP), respectively. In the freshwater ponds the values were 0.63 and 0.79 for the freshwater-duckweed-fed pond (FDP) and freshwater-commercial feed-fed pond (FCP), respectively. Extrapolation of results revealed that the fishponds fed on fresh duckweed provided a net fish yield of 15.7 ton/ha/y in the TDP and 8.5 ton/ha/y in the FDP. In case of commercial feed-fed ponds the net fish yields were, 16.8 ton/ha/y and 11.2 ton/ha/y in the TCP and FCP, respectively. The feed conversion ratios (FCRs) were 1.2, 1.4, 1.6 and 1.5 in, TDP, TCP, FDP and FCP, respectively. The N balance in the fish ponds indicated that, the percentages of the nitrogen recovery from the total nitrogen input, represented 13.4% and 13.7% in treated sewage fed ponds, and, 24.2% and 27.9% in the freshwater fed ponds. The results show that reuse of treated sewage and sewage-grown duckweed in aquaculture produces high fish yield and efficient nutrients recovery at low feeding costs.

Keywords: Reuse; Treated domestic sewage; Fresh duckweed; *Lemna gibba; Oreochromis niloticus;* UASB-duckweed ponds

1. Introduction

Fish is an important and cheap source of animal protein in human diets. Three quarters of the world catch is used for direct human consumption and the rest for fishmeal and fish oil production (FAO, 2000). Inland captures provide a significant proportion of the world catch and form an important source of animal protein in some countries (FAO, 2000). However, the harvest of fish from rivers and lakes suffers from environmental degradation and other uses of the aquatic ecosystem. Under such circumstances, the culture of fish may constitute an interesting alternative for the declining catches. Fish aquaculture has increased by 58% during the period from 1994 to 1999, while captured fish remained stable (FAO, 2000). Inland aquaculture accounted for 60% of the world aquaculture production. Since fish aquaculture is the only sustainable way to fill the gap between supply and demand for fisheries products, a global strategy was established by FAO and other international, regional and national organisations to increase its contribution in food security.

Water is one of the most precious elements in fish farming systems, especially in water scarce countries. In arid and semiarid countries like Egypt, more than 80% of the water consumption is being used for crop irrigation. In these countries competition for water exists between fish culture and crop irrigation. Governmental authorities usually give priority to the latter. However, fishponds could be fed with treated domestic wastewater representing an alternative source of water. Another important element in fish farming is the fertilisation of ponds by either using feed or fertilisers. In developed countries high nutritional feed pellets, which are not readily available in most developing countries, are being used. On the other hand, in most developing countries, organic waste and agricultural products, by-products and wild plants have been used as fish feed (El-Sayed, 1992; Keke *et al.*, 1994; Basudha and Vishwanath, 1997; Mbahinzireki *et al.*, 2001; Middleten *et al.*, 2001), making this type of farming more accessible for local resources and social groups with limited economic resources. Allela (1985) reported that improvement of aquaculture in Kenya is depending upon the availability of suitable and cheap supplementary feed, using thereby local agricultural products or organic waste products.

The current study focuses on the use of duckweed biomass grown on anaerobically pre-treated domestic sewage as well as the effluent from duckweed sewage stabilisation ponds in fish farming. The use of duckweed (*Lemna gibba*) as a sole fish feed was compared to locally available commercial fish feed. The use of treated domestic sewage in fish aquaculture was compared with freshwater (de-chlorinated city tap water) aquaculture.

2. Materials and methods

2.1. Fish species

Experiments were carried out at the Water Pollution Control Department, National Research Centre, Cairo; Egypt. Juveniles of Nile tilapia (*Oreochromis niloticus*) with mean body weight of 20 grams were bought from a local commercial fish farm. Two weeks before the experiment, the fish were adapted to the environmental conditions in one holding tank with an effective water volume of 480 l and 48 cm depth. De-chlorinated and continuously aerated tap water was supplied to the tank at a rate of 50 l/day. Continuous aeration of the tank was performed using a compressor. The juveniles were fed to satiation on a mixture of fresh duckweed harvested from duckweed ponds, treating anaerobically (UASB) pre-treated domestic sewage and fishmeal-based commercial fish feed. Continuous removal of uneaten feed and faeces was performed daily during the adaptation period.

2.2. Experimental design

Operating conditions are presented in Table 1. At the start of the experiment, 40 juveniles from the holding tank were randomly distributed between four tanks, which were previously filled with either treated sewage (tanks 1 and 2) or tap water (tanks 3 and 4) at a stocking density of 10 fish per tank. The total nitrogen content of each pond was determined before stocking the fish. Each tank had 1 m^2 surface area and 480 l working volume.

Table 1: Operating conditions in commercial feed and duckweed fed fish ponds

Item	TDP	TCP	FDP	FCP
Fish weight (g fish^{-1})	23.27±3.80	23.18±3.78	24.60±3.91	22.58±3.00
Feed source	Fresh duckweed	Commercial feed	Fresh duckweed	Commercial feed
Feeding rate (g fresh feed/100 g fish)	25	1.33-1.40	25	1.33-1.40
Water source	Treated sewage	Treated sewage	Freshwater	Freshwater
Flow rate (l d^{-1})	44.5	44.5	45	45

Table 2: Characteristics of the treated effluent used in TDP and TCP

Parameters	Treated sewage
Temperature °C	25-28.5
pH	6.9-7.4
COD (mg O_2/l)	83±35
BOD (mg O_2/l)	16±5
Total ammonia nitrogen (mg N/l)	3.2±1.9
Nitrite nitrogen (mg N/l)	0.35±0.29
Nitrate nitrogen (mg N/l)	0.61±0.5
TKN (mg N/l)	8.5±1.3
TP (mg P/l)	1.66±0.42
TSS (mg /l)	44±15
DO (mg O_2/l)	6.4±0.6

2.3. Water sources and fish feed

Treated domestic sewage from a pilot treatment plant consisting of a UASB reactor followed by duckweed ponds (El-Shafai *et al.*, submitted) and de-chlorinated tap water were used as water sources. Characteristics of the treated sewage are presented in Table 2. Effluent from the treatment system was fed to 2 fishponds at an average daily flow rate of 44.5 l each. Two additional fishponds were fed with de-chlorinated tap water at a daily flow rate of 45 l each. Sources of feed that were compared included fresh duckweed (*Lemna gibba*) harvested from pilot-scale duckweed stabilisation ponds, and a fishmeal-based commercial fish feed. The ingredients of the latter feed were fishmeal, fish oil, gluten, wheat bran and vitamins. The composition of duckweed and commercial feed is presented in Table 3. Feeding rate in the duckweed-fed fishponds was 25 g fresh weight per 100 g fish live body weight per day. Tilapia in ponds fed commercial feed received dry feed at a rate ensuring equal amount of protein addition as for the duckweed-fed ponds (iso-nitrogenous diets). Commercial feed was added to ponds in a small feeding tray to avoid feed loss in the pond. Fishponds were named treated sewage-duckweed-fed pond (TDP), treated sewage-commercial feed-fed pond (TCP), freshwater-duckweed-fed pond (FDP) and freshwater-commercial feed-fed pond (FCP).

Table 3: Quality of duckweed and commercial feed

Parameters	Duckweed	Commercial feed
Dry matter (%)	4.5±0.5	92.7±2.5
Protein content (%)[1]	37± 2.6	32.9±1.2

[1] Based on dry matter

2.4. Monitoring and analytical procedures

The experiment lasted 120 days, during which monitoring of the water in the ponds and of the treated sewage was performed. Samples were collected once a week for analysis of, chemical oxygen demand (COD), biological oxygen demand (BOD), total suspended solids (TSS), nitrate nitrogen, TKN and total phosphorus (TP). Temperature, pH, total ammonia nitrogen and nitrite nitrogen were analysed twice a week. All the analyses were carried out according to Standard Methods (APHA, 1998). During the study period evaporation rate was measured using miniponds. The fish in each treatment was weighed individually once every month and specific growth rates were calculated based on the following expression:

$$SGR = \frac{(\ln Wf - \ln Wi)}{days} \times 100 \tag{1}$$

Where, Wi and Wf are initial and final mean body weight respectively.

The total feed intake and total protein intake were recorded in each treatment and by the end of the trial the feed conversion ratio (FCR) and protein efficiency ratio (PER) were calculated based on the following expressions:

$$FCR = \frac{total\ feed\ ingested\ (dry\ weight)}{fish\ weight\ gain\ (wet\ weight)} \tag{2}$$

$$PER = \frac{fish\ weight\ gain\ (wet\ weight)}{total\ protein\ ingested} \tag{3}$$

Accumulated sediment in each treatment was collected at the end of the trial for analyses. The nitrogen content of the sediment was included in the nitrogen mass balance. Fish mortality, total fish yield and net fish yield in ton/ha/year and nitrogen mass balance were defined in each treatment.

$$\text{Total fish yield} = \frac{(\text{Final crop density in ton / ha})(365)}{\text{Growth period in days}} \qquad (4)$$

$$\text{Net fish yield} = \frac{(\text{Initial density in ton / ha} - \text{Final density in ton / ha})(365)}{\text{Growth period in days}} \qquad (5)$$

2.5. Statistical analysis

Results of SGRs are presented as mean values plus or minus standard deviations. Specific growth rates (SGRs) in each treatment were subjected to one-way analysis of variance (one-way ANOVA). P value of 5% was chosen for the non-significant difference and 1% for the significant difference.

Table 4: Water quality parameters in the fishponds

Parameter	TDP	TCP	FDP	FCP
Temperature °C	27-34	27-34	26-32	26-32
pH	8.6-9.1	8.3-9.1	8.7-9.9	8.6–9.6
COD (mg O_2/l)	330±81	248±56	213±19	226±26
BOD (mg O_2/l)	58±13	42±9	40±3	42±5
TAN[1] (mg N/l)	ND[2]	0.02±0.07	ND[2]	ND[2]
Nitrite(mg N/l)	0.022±0.005	0.026±0.010	0.010±0.001	0.011±0.001
Nitrate (mg N/l)	0.18±0.04	0.15±0.054	0.16±0.06	0.14±0.05
TKN (mg N/l)	13.8±3.5	14.5±3.8	3.2±0.7	3.5±1
TP (mg P/l)	1.85±0.31	1.29±0.4	0.46±0.17	0.32±0.09
TSS (mg /l)	213± 49	165±38	162±13	150±17
DO (mg O_2/l)	13±1.3	13.9±1.6	14.2±1.6	14.6±1.2

[1] TAN total ammonia nitrogen
[2] ND not detected

3. Results

3.1. Water quality

The water quality remained good for all fishponds throughout the experimental period. There appeared to be no accumulation of toxic metabolic by-products of the fish such as ammonia or nitrite (Table 4). The temperature range was 26-34 °C in all ponds. The oxygen concentration in all ponds was around 14 mg O_2/l probably due to algal activity.

3.2. Growth performance

Growth performance was assessed by calculating, SGR, FCR, PER and (total and net) fish yield (Table 5). The results of the SGR showed no significant ($p > 0.05$) difference between juveniles of tilapia fed either on duckweed or commercial feed when using treated sewage as the sole water source. In case of freshwater ponds, the SGR of tilapia fed on commercial feed was significantly ($p < 0.01$) higher than SGR of fish fed on duckweed. In case of fresh duckweed treatments, the SGR of tilapia reared in treated sewage was significantly ($p < 0.01$) higher than SGR of tilapia reared in freshwater. Also in case of commercial feed treatments, SGR was higher in the pond receiving treated effluent ($p < 0.01$) than in the freshwater pond. The feed conversion ratios (FCRs) were 1.2, 1.4, 1.6 and 1.6 in TDP, TCP, FDP and FCP, respectively. The protein efficiency ratios were, 2.28, 2.17, 1.7 and 2.04, respectively.

Table 5: Growth performance in fishponds

Parameter	TDP	TCP	FDP	FCP
Initial mean body weight (g fish^{-1})	23.27±3.80	23.18±3.78	24.60±3.91	22.58±3.00
Final mean body weight (g fish^{-1})	74.99±15.75	78.46±14.04	52.39±9.83	59.30±14.95
Weight gain (g fish^{-1})	51.72±12.23	55.28±10.42	27.79±6.77	36.72±12.10
Feed input (g dry feed fish^{-1})	62.78	77.29	44.76	54.73
SGR*	0.97±0.08^a	1.01±0.08^a	0.63±0.04^b	0.79±0.11^c
FCR	1.2	1.4	1.6	1.5
PER	2.28	2.17	1.7	2.04

*$p > 0.05$ non significant different, Values with similar superscript letter (a, b or c) are non significantly different

*$p < 0.01$ significant different, Values with different superscript letter (a, b or c) are significantly different

Total fish yields were 22.8, 23.9, 15.9 and 18 ton/ha/y in TDP, TCP, FDP and FCP, respectively while net fish yields were, 15.7, 16.8, 8.5 and 11.2 ton/ha/y, respectively (Fig. 1).

3.3. Nitrogen mass balance

Nitrogen mass balances in the ponds showed that the percentages of nitrogen recovery in fish from the total nitrogen input were, 13.4%, 13.7%, 24.2% and 27.9% in TDP, TCP, FDP and FCP, respectively (Table 6). A major part of added nitrogen left the pond system via the effluent, mainly as organic nitrogen (algae). The percentages of the total nitrogen input discharged with the effluents were, 80.7%, 80%, 66.4% and 64.9% in the TDP, TCP, FDP and FCP, respectively.

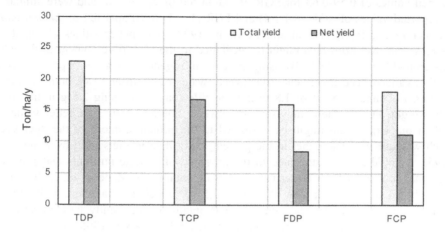

Fig. 1: Total and net fish yields in fishponds in ton/ha/y

Table 6: Nitrogen mass balance in fishponds in g/tank

Item[**]	TDP	TCP	FDP	FCP
$TN_{input.}$ (g tank^{-1})	85.76	90.1	26.05	28.8
$TN_{reco.}$ (g tank^{-1})	11.52 (13.4%)[*]	12.3 (13.7%)[*]	6.3 (24.2%)[*]	8.04 (27.9%)[*]
$TN_{disc.}$ (g tank^{-1})	69.24 (80.7%)[*]	72.15 (80%)[*]	17.3 (66.4%)[*]	18.68 (64.9%)[*]
$TN_{sed.}$ (g tank^{-1})	4.7 (5.5%)[*]	5 (5.6%)[*]	2 (7.7%)[*]	1.8 (6.2%)[*]
$TN_{unacc.}$ (g tank^{-1})	0.3 (0.4%)[*]	0.65 (0.7%)[*]	0.45 (1.7%)[*]	0.28 (1%)[*]

[*] Values between parentheses are % from the total nitrogen input.

[**] $TN_{input.}$ total nitrogen input, $TN_{reco.}$ total nitrogen recovered in fish, $TN_{disc.}$ total nitrogen discharged via effluent, $TN_{sed.}$ total nitrogen in the sediment and $TN_{unacc.}$ Unaccounted nitrogen (all in gram)

4. Discussion

4.1. Growth performance

In the different treatment tanks the range of SGR and FCR varied between 0.63 and 1.01 and between 1.2 and 1.6, respectively. These ranges are comparable to published data for Nile tilapia and other fish species using different feed pellets (Table 7). In duckweed fed ponds the SGR was 0.97 and 0.63 in TDP and FDP while FCR was 1.2 and 1.6, respectively. These values of SGR are similar to the value of 0.8 obtained by Fasakin *et al.* (1999) for duckweed-based diet fed tilapia. FCR in our study is far better than value of 4.3 reported by Fasakin *et al.* (1999). Our results are also better than the reported values of 0.59-0.63 for SGR of tilapia fed on *Spirodella* and were similar to the reported values of SGR (0.4-1.4) and FCR (1-3.3) for tilapia fed on *Lemna gibba* (Gaigher *et al.*, 1984; Hassan and Edwards, 1992). The better values obtained for *Lemna gibba* over *Spirodela* might be attributed to the high fibre content in the latter (Hassan and Edwards, 1992). By using three iso-nitrogenous diets (28%) prepared from locally available feed ingredients (duckweed, water hyacinth and rice bran), Zaher *et al.* (1995) obtained a SGR of 1.8 to 2.2 for Nile tilapia weighing 2.5 g initial mean body weight. FCR in their study ranged between 2.09 and 2.46. The better values of FCR in our experiment might be attributed to the protein content of the duckweed, which was almost 10% higher in the present study than in diets used by Zaher *et al* (1995). It could also be attributed to the presence of sewage nutrients that provided additional sources of feed. Better values of SGR obtained by Zaher *et al.* (1995) could be attributed to the small size of tilapia (2.5 g), which is characterised by higher metabolic and growth rate. The values of FCR obtained in this study are comparable to FCR (1.05-1.33) obtained in freshwater fish farms in Denmark in 1995 (Iversen, 1995).

The results of PER in the range of 1.7-2.28 obtained in this study are better than values of Nile tilapia fed on diets containing *Spirodela* (Fasakin *et al.*, 1999) and tilapia fed on diets containing cocoa husk (Falaye and Jauncey, 1999). The poor results in case of *Spirodela* and cocoa hush in comparison to *Lemna gibba* might be attributed to the high fibre content and low protein content of *Spirodela* and cocoa husk. The higher fibre content negatively affects protein digestibility and growth rate. Our range is better than the reported range (1.1-1.8) for tilapia fed on formulated feed containing palm kernel oil as substitute of soybean meal (Lim *et al.*, 2001) and Nile tilapia fed on plant protein-based diet (Fontainhas *et al.*, 1999). This is probably attributed to the negative effect of *Asprgillus flavus* contained in palm kernel meal (Lim *et al.*, 2001) and anti-nutritional factors in the plant ingredients used by Fontainhas *et al.* (1999). On the other hand our results are lower than reported values (2.46-2.87) of tilapia fed on formulated feed containing barely seed as substitute of corn meal (Belal, 1999). The advantage of the diet used by Belal could be attributed to the experimental set-up since he used a recirculation system. It also might be attributed to some growth factors and/or digestible energy derived from fishmeal in his diet.

Table 7: SGR and FCR of some fish species

Fish species	Fish feed	F W[1]	SGR	FCR	References
Nile tilapia	Pellets	5	0.7	N D[2]	Keke *et al.*, 1994
Nile tilapia	Pellets	2.5	2.2	2.1	Basudha and Vishwanath, 1997
Blue tilapia	Pellets	27	1.2	N D[2]	Papoutsoglau and Tziha, 1996
Hybrid tilapia	Pellets	68	0.9	N D[2]	Middleton *et al.*, 2001
Nile tilapia	Pellets	15	1.4	1.6	Mbahinzireki *et al.*, 2001
Nile tilapia	Pellets	40	1.13	2.1	El-Sayed, 1992
Rainbow trout	Pellets	N D[2]	0.7-1	1.1-1.6	Thorpe and Choc, 1995
Rainbow trout	Pellets	N D[2]	1.8-2.1	N D[2]	Cowey, 1995
Atlantic salmon	Pellets	N D[2]	0.7-1	1.3-1.4	Thorpe and Choc, 1995

[1] F W average fish weight

[2] N D means no data available

4.2. Fish yield

The net annual fish yield was in the range of 8.5-16.8 ton/ha/y in the treatment tanks and provided 15.7 and 8.5 ton/ha/y in TDP and FDP, respectively (Fig. 1). The yield is better than the net fish yield (1.5 ton/ ha/y) of Indian major carps reared in ponds-fed with cattle-shed waste, biogas slurry and water hyacinth (Mishra *et al.*, 1988). High yields in our study are mainly attributed to the good quality of treated sewage as a pond fertiliser in comparison to cattle manure, which has been defined as a poor quality pond fertiliser (Edwards *et al.*, 1994) and to the high protein and low fibre content of duckweed in comparison to water hyacinth. The maximum achievable yield of tilapia, snakehead and milkfish from fertilised pond supplied with supplemental feed, rice bran or copra meal was 2.4 ton/ha/y (Cruz and Laudencia, 1980). This low yield might be attributed to low protein content of rice bran and copra meal and low stocking density (6000-8000 fingerling/ha) in comparison to our study (100,000 fingerling/ha). In Bangladesh, annual net fish yield of carp polyculture ponds supplied with fresh duckweed was in the range of, 2.7-6.28 ton/ha/y (Azim and Wahab, 1998; Wahab *et al.*, 2001; Azim *et al.*, 1998). Lower fish yields in these experiments could be attributed to the low stocking density (10000-15000 fingerling/ha) applied. Our results are better than the net fish yield of milkfish ponds fertilised with inorganic fertiliser, cow manure or duckweed as reported by Ogburn and Ogburn (1994). In the latter study, the maximal fish yield was obtained in ponds fertilised with duckweed, achieving 3.3 ton/ha/y in comparison to 2.2 ton/ha/y in organic fertilised ponds and 1.3 ton/ha/y in inorganic fertilised ponds. A major difference in the Ogburn and Ogburn (1994) study was that the duckweed was used in a dry form as a pond fertiliser before stocking the pond with fish, while in our study the duckweed was used as a direct fish feed, alone or in addition to treated sewage as pond fertiliser. The production of grass carp in outdoor cement tanks supplied with Napier grass (*Pennisetum pupureum*) as the sole pond input provided 5.9 ton/ha/y net fish yield (Shrestha, 1999). At Sagana fish farm in central Kenya, polyculture of Nile tilapia with African catfish has been conducted in ponds fed with chemical fertilisers, rice bran or combinations of both (Omondi *et al.*, 2001). The best net fish yield about 4 ton/ha/y was obtained in the chemical fertiliser-rice bran fed ponds. The results show the superior quality of duckweed over rice bran, and advantages of treated sewage (organic manure) over inorganic fertilisers. Tacon *et al.*, (1995) reported that attainable fish yield that could be harvested from semi-intensive pond supplied with supplemental feed and pond fertilisers in the range of 1-15 ton/ha/y. Our results of annual fish yields are close to the upper limit.

In our study the treated sewage-fed fishponds, provided better results than the freshwater-fed ponds. Cruz and Laudencia (1980) reported that the fish yield from fertilised ponds supplied with supplemental feed was significantly higher than those without supplemental feed. Under tropical and sub-tropical conditions where primary and secondary productivity is rather high, the contribution of natural aquatic organisms to the nutrition of farmed fish could be significant. In organically fertilised ponds of freshwater prawns, *Macrobracium rosenbergii*, natural food organisms, especially cladocera, copepods and oligochaets, contributed 15% to 43% to the growth of prawns (Tidwell *et al.*, 1995; Augustin, 1999).

The low growth performance and net yield, in FDP compared to FCP might be attributed to deficiency of some essential elements and additives in duckweed like cysteine and methionine (Hammouda *et al.* 1995). These essential elements might be supplied by the natural food organisms in case of the treated sewage fed pond, resulting in no significant differences between the duckweed and commercial feed. This possible role of natural food has been reported earlier by D'Abramo and Conklin (1995) who stated that natural biota could significantly contribute to the requirements of nutrients, such as vitamins, cholesterol, phospholipids and minerals (D'Abramo and Conklin, 1995; Milstein *et al.*, 1995; Omondi *et al.*, 2001). The treated effluent in this experiment provided a good food source for tilapia in addition to the supplemented feed. Biomass of daphnia and copepods, have been detected in the treated effluent (El-Shafai *et al.*, submitted). This daphnia has excellent feed value for pisciculture considering the protein content (59.5% dry matter) and high-energy content (20 KJ g-1 dry matter), which are among the highest values mentioned in the literature (Bogatova *et al.*, 1973; Proulx and de La Noue, 1985).

4.3. Nitrogen mass balance

The results of the N mass balance showed that in treated sewage-fed ponds about 13% of the nitrogen input, has been recovered by fish and transformed into fish protein compared to 24-28% for the freshwater-fed ponds. A possible explanation for this result might be that the dissolved Nitrogen in the treated sewage was not available for fish, leading to higher N discharges in the treated sewage fed ponds than in freshwater fed ponds. Literature data shows that in semi-intensive farms based on supplemental feed input, nitrogen recovery in fish ranged from 5% to 25% (Tacon *et al.*, 1995). The nitrogen recovery in the treated sewage fed ponds is similar to the reported values of 11% to 16% in polyculture freshwater fishpond receiving organic manure or chemical fertiliser (Schroeder *et al.*, 1990). The results are comparable to reported values (15.5-21%) in chicken litter tilapia fertilised pond, diet-based channel catfish pond and diet-based *Striped bass* brackish water pond (Worsham 1975; Daniels and Boyd, 1989; Green and Boyd, 1995). These results showed a good digestibility of the duckweed protein and a good nutritional value as sole fish feed. In this study the N recovery in freshwater-fed ponds (24-28%) is comparable to the reported values (30-36%) for *Sparus aurata* (Krom *et al.*, 1985; Porter *et al.*, 1987) and 27% for channel catfish (Boyd 1985). Gaigher *et al* (1984) reported that about 49% of the nitrogen content of fresh duckweed, was recovered by Nile Tilapia. The higher value of Gaigher could be attributed to the optimal growth condition since he used a re-circulation system with optimal conditions during 24 hours a day. Gomes *et al* (1995) reported that the nitrogen retention represented 15.5-30% from the dietary nitrogen in trout growth trials using a fishmeal based diet and a diet containing plant ingredients as replacement of fishmeal. while Phillips and Beveridge (1986) reported 21% nitrogen recovery.

In this study the nitrogen content of the pond sediment accounted for 5.5-7.7% from the total nitrogen input, which lies within the range of 4.3% and 13.7%, reported by Gomes *et al* (1995). The range is also similar to the average value (6%) reported in the sediment of shrimp pond and suggests rapid mineralisation of organic waste by the microbial communities in the sediment. This results in diffusion of ammonia into the water making it available to the phytoplankton (Burford *et al.*, 2002). Gaigher *et al* (1984) reported that 14% of nitrogen input of fresh *Lemna gibba* accumulated in the sediment of Nile tilapia pond. This higher value of faecal nitrogen could be attributed to daily remove of faeces so low biodegradation and mineralisation rate was occurred. In polyculture ponds Liang *et al.* (1999) reported that 0.53% to 2% from the total nitrogen input accumulated in the sediment. This low value could be attributed to high mineralisation since Liang *et al* (1999) reported also a low DO and high ammonia nitrogen in their ponds.

The major part of discharged nitrogen in the pond effluent is organic nitrogen (Table 4). This suggests that the principal sink of ammonia in the pond water is its uptake by phytoplankton and that the loss by ammonia volatilisation is minor (Hargreaves, 1998). The unaccounted part (Table 6) is proposed to be ammonia volatilisation and/or nitrogen denitrification. Despite the large volume of sediment, the magnitude of N removal by denitrification is low because both the nitrification and denitrification are tightly coupled in aquatic sediments where nitrification is limited by the depth of oxygen penetration in the sediment (Hargreaves, 1998). The kinetic of nitrification is first order with respect to substrate (ammonia) concentration so low concentration of ammonia due to phytoplankton uptake imposes substrate limitation on the nitrification in ponds (Hargreaves, 1998).

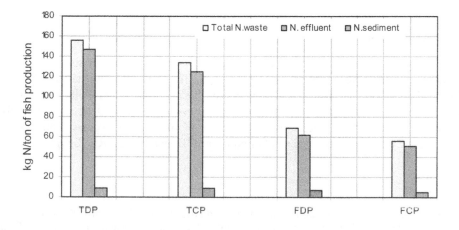

Fig. 2: Nitrogen wastes in fish tanks as kg N/ton of fish production

4.4. Nitrogen wastes

Protein is the major limiting factor in fish feed and the price of feed is mainly controlled by quantity and quality of dietary protein. Use of good quality protein reduces the nitrogen loss in the pond, which represent money save and prevent quality deterioration of the pond water. The total nitrogen discharge from the ponds ranged between 56 and 156 Kg N.ton fish^{-1}, Fig. 2. The range for the freshwater fed ponds was 56-69.5 KgN.ton fish^{-1}, which is similar to the amount of nitrogen (53-71 Kg N) released by the production of one ton of carp or by one ton of shrimp production (26-117 Kg N) (Tacon *et al.*, 1995). In typical Nordic freshwater fish farms the nitrogen release to the environment was estimated to be 132 kg N in 1974 and 55 Kg N in 1995 for each ton of fish production (Enell, 1995) while the value is 125 Kg N in the North Ireland fish farm (Foy and Rosell, 1991a; Foy and Rosell, 1991b). Variation between the published data might be attributed to the quality of the feed and its protein digestibility, fish species and more strongly to the N load and operating conditions.

5. Conclusions

-The fresh duckweed-tilapia production is applicable in pond system, technically and economically than other organic waste fed fish farming systems.

-The treated sewage fertilised ponds supplied with fresh duckweed is superior to the non-fertilised pond and similar to commercial feed-fed pond supplied with freshwater.

-Fresh duckweed (*Leman gibba*) is more suitable for small-scale fish farmers in rural communities than high price fish feed.

-There is no any preference for the commercial feed over the fresh duckweed in tilapia ponds fertilised with treated sewage.

-Fertilisation of fishpond fed with commercial feed, using treated sewage increases pond productivity and fish yield.

-Further research is needed to optimise nitrogen recovery at higher feeding rates

Acknowledgements

The authors would like to thank the Dutch government for financial support for this research within the framework of the SAIL funded "Wasteval" project (LUW/MEA/971). The project is a co-operation between the Water Pollution Control Department (NRC, Egypt), Wageningen University and UNESCO-IHE Institute for Water Education, The Netherlands. The authors would like to express their sincere thanks to the technicians of Water Pollution Control Department, NRC, Egypt.

References

Agustin, C.P., 1999. Role of natural productivity and artificial feed in the growth of freshwater prawn, *Macrobrachium borelli* (Nobili, 1896) cultured in enclosures. J. Aquacult. Trop. 14(1), 47-56.

Allela, S.D., 1985. Development and research of aquaculture in Kenya. In: (Eds.), Aquaculture Research in The African Region "Proceeding of The African Seminar on Aquaculture, organized by the IFS, 7-11 October 1985, pp. 112-114.

APHA, 1998. Standard Methods for the Examination of Water and Wastewater, 20th edition, Washington.

Azim, M.E. and Wahab, M.A., 1998. Effects of duckweed (*Lemna sp.*) on pond ecology and fish production in carp polyculture of Bangladesh. Bangladesh J. Fish. 21(1), 17-28.

Azim, M.E., Wahab, M.A., Haque, M.M., Wahid, M.I. and Haq, M.S., 1998. Suitability of duckweed (*Lemna sp.*) as dietary supplement in four species polyculture. Progress. Agric. 9(1 & 2), 263-269.

Basudha, C. and Vishwanath, W., 1997. Formulated feed based on aquatic weed azolla and fishmeal for rearing carp *Osteobrama belangeri* (Valenciennes). J. Aquacult. Trop. 12, 155-164.

Belal, I.E.H., 1999. Replacing dietary corn with barley seed in Nile tilapia, *Oreochromis niloticus* (L.), feed. Aquacult. Res. 30, 265-269.

Bogatova, I.B., Schhesbina, M.A., Ovinnikova, V.V. and Tagisova, N.A., 1973. The chemical composition of certain planktonic animals under different growing conditions. Hydrobiol. J. 7, 39-43.

Boyd, C.E., 1985. Chemical budgets of channel catfish ponds. Trans. Am. Fish. Soc. 114, 291-298.

Burford, M.A., Preston, N.P., Gilbert, P.M. and Dennison, W.C., 2002. Tracing the fate of ^{15}N-enriched feed in an intensive shrimp system. Aquaculture 206, 199-216.

Cowey, C.B., 1995. Intermediary metabolism in fish with reference to output of end products of nitrogen and phosphorus. Water. Sci. Technol. 31(10), 21-28.

Cruz, E.M. and Laudencia, L., 1980. Polyculture of milkfish (Chanos chanos Forskal), all male Nile tilapia (*Tilapia nilotica*) and snakehead (*Ophicephalus striatus*) in freshwater ponds with supplemental feeding. Aquaculture 20, 231-237.

D'Abramo, L.R. and Conklin, D.E., 1995. New development in the understanding of the nutrition of penaeid and caridean species of shrimp. Proceedings special session on shrimp farming, aquaculture 95, World Aquaculture Society, Louisiana, USA, pp. 95-107

Daniels, H.V. and Boyd, C.E., 1989. Chemical budgets for polyethylene line, brackish water ponds. J. World Aquacult. Soc. 20(2), 53-59.

Edwards, P., Kaewpaitoon, K., Little, D.C. and Siripandh, N., 1994. An assessment of the role of buffalo manure for pond culture of tilapia. II: field trial. Aquaculture 126, 97-106.

El-Sayed, A.F.M., 1992. Effects of substituting fish meal with *Azolla pinnata* in practical diets for fingerling and adult Nile tilapia, *Oreochromis niloticus* L. Aquacult. Fish. Manag. 23, 167-173.

Enell, M., 1995. Environmental impact of nutrients from Nordic fish farming. Water Sci. Technol. 31(10), 61-71.

Falaye, A.E. and Jauncey, K.C., 1999. Acceptability and digestibility by tilapia Oreochromis niloticus of feed containing cocoa husk. Aquacult. Nutr. 5, 157-161.

FAO, 2000. The state of the world fisheries and aquaculture, edited by editorial group, FAO information division.

Fasakin, E.A., Balogun, A.M. and Fasuru, B.E., 1999. Use of duckweed, *Spirodela polyrrhiza* L. schleiden, as a protein feed stuff in practical diets for tilapia, *Oreochromis niloticus* L. Aquacult. Res. 30, 313-318.

Fontainhas, F.A., Gomes, E., Reis-Henriques, M.A., and Coimbra, J., 1999. Replacement of fishmeal by plant proteins in the diet of Nile tilapia: digestibility and growth performance. Aquacult. International 7, 57-67.

Foy, R.H. and Rosell, R., 1991a. Fractionation of phosphorus and nitrogen loading from a Northern Ireland fish farm. Aquaculture 96, 31-42.

Foy, R.H. and Rosell, R., 1991b. Loading of nitrogen and phosphorus from a Northern Ireland fish farm. Aquaculture 96, 17-30.

Gaigher, I.G., Porath, D. and Granoth, G., 1984. Evaluation of duckweed (*Lemna gibba*) as feed for tilapia (*Oreochromis niloticus* × *Oreochromis aureus*) in a recirculating unit. Aquaculture 41, 235-244.

Gomes, E.F., Rema, P., Gouveia, A. and Teles, A.O., 1995. Replacement of fishmeal by plant proteins in diets for Rainbow trout (*Oncorehynchus mykiss*): effect of quality of fishmeal based control diets on digestibility and nutrient balances. Water Sci. Technol. 31(10), 205-211.

Green, B.W. and Boyd, C.E., 1995. Chemical budgets for organically fertilised fishponds in the dry tropics. J. World Aquacult. Soc. 26(3), 284-296.

Hammouda, O., Gaber, A. and Abdel-Hameed, M.S., 1995. Assessment of the effectiveness of treatment of wastewater-contaminated aquatic systems with *Lemna gibba*. Enz. and Micro. Technol. 17, 317-323.

Hargreaves, J.A., 1998. Nitrogen biogeochemistry of aquaculture ponds. Aquaculture 166, 181-212.

Hassan, M.S. and Edwards, P., 1992. Evaluation of duckweed (*Lemna perpusilla* and *Spirodela polyrrhiza*) as feed for Nile tilapia (*Oreochromis niloticus*). Aquaculture 104, 315-326.

Iversen, T.M., 1995. Fish farming in Denmark: Environmental impact of regulative legislation. Water Sci. Technol. 31(10), 73-84.

Keke, I.R., Ofajekwu, C.P, Ufodike, E.B. and Asala, G.N., 1994. The effect of partial substitution of groundnut cake by water hyacinth (*Eichhornia crassipes*) on growth and food utilisation in the Nile tilapia, *Oreochromis niloticus* (L.). Acta Hydrobiol. 36(2), 235-244.

Krom, M.D., Porter, C. and Gordin, H., 1985. Nutrient budget of a marine fishpond in Eilate, Israel. Aquaculture 51, Pp. 65-80.

Liang, Y., Cheung, R.Y.H., Everitt, S. and Wong, M.H., 1999. Reclamation of wastewater for polyculture of freshwater fish: Fish culture in ponds. Water Res. 33(9), 2099-2109.

Lim, H.A., Ng, W.K., Lim, S.L. and Ibrahim, C.O., 2001. Contamination of palm kernel meal with *Aspergillus flavus* affects its nutritive value in pelted feed for tilapia, *Oreochromis mossambicus*. Aquacult. Res. 32, 895-905.

Mbahinzireki, G.B., Dabrowski, K., Lee K.J., El-Saidy, D. and Wisner, E.R., 2001. Growth, feed utilisation and body composition of tilapia (*Oreochromis sp.*) fed with cottonseed meal-based diets in a recirculation system. Aquacult. Nutr. 7, 189-200.

Middleton, T.F., Fekert, P.R., Boyd, L.C., Daniels, H.V. and Gallagher, M.L., 2001. An evaluation of co-extruded poultry silage and culled jewel sweet potatoes as a feed ingredient for hybrid tilapia (*Oreochromis niloticus* × *Oreochromis mossambicus*). Aquaculture 198, 269-280.

Milstein, A., Alkon, A. and Karplus, I., 1995. Combined effects of fertilisation rate, manuring and feed pellet application on fish performance and water quality in polyculture ponds. Aquacult. Res. 26, 55-65.

Mishra, B.K., Sahu, A.K. and Pani, K.C., 1988. Recycling of aquatic weed, water hyacinth, and animal wastes in rearing of Indian major carps. Aquaculture 68, 59-64.

Ogburn, D.M. and Ogburn, N.J., 1994 Use of duckweed (lemna spp.) grown in sugar mill effluent for milkfish, *Chanos chanos* Forskal, production. Aquacult. Fish. Manag. 25, 497-503.

Omondi, J.G., Gichuri, W.M. and Veverica, K., 2001. A partial economic analysis for Nile tilapia (*Oreochromis niloticus* L.) and sharptoothed catfish *Clarias gariepinus* (Burchell 1822) polyculture in central Kenya. Aquacult. Res. 32, 693-700.

Papoutsoglou, S.E. and Tziha, G., 1996. Blue tilapia (*Oreochromis aureus*) growth rate in relation to dissolved oxygen concentration under re-circulated water conditions. Aquacult. Eng. 15(3), 181-192.

Phillips, M. and Beveridge, M., 1986. Cages and the effect on water condition. Fish Farmer 9(3), 17-19.

Porter, C.B., Krom, M.D., Robbins, M.G., Brickell, L. and Davidson, A., 1987. Ammonia excretion and total N budget for gilthead seabream (*Sparus aurata*) and its effect on water quality conditions. Aquaculture 66, 287-297.

Proulx, D. and De La Noue, J., 1985. Harvesting *Daphnia magna* grown on urban tertiarily treated effluents. Water Res. 19(10), 1319-1324.

Schroeder, G.L., Wohlfarth, G., Alkon, A., Halevy, A. and Krueger, H., 1990. The dominance of algal-based food webs in fishponds receiving chemical fertiliser plus organic manure. Aquaculture 86, 219-229.

Shrestha, M.K., 1999. Summer and winter growth of grass carp (*Ctenopharyngodon idella*) in a poly culture fed with Napier grass (*Pennisetum purpureum*) in the subtropical climate of Nepal. J. Aquacult. Trop. 14(1), 57-64.

Tacon, A.G.J., Phillips, M.J. and Barg, U.C., 1995. Aquaculture feeds and the environment: the Asian experience. Water Sci. Technol. 31(10), 41-59.

Thorpe, J.E. and Cho, C.Y., 1995. Minimising waste through bioenergetically and behaviourally based feeding strategies. Water Sci. Technol. 31(10), 29-40.

Tidwell, J.H., Webster, C.D., Sedlacek, J.D., Weston, P.A., Knight, W.L., Hill, Jr S.J., D'Abramo, L.R., Daniels, W.H., Fuller, M.J. and Montanez, J.L., 1995. Effects of complete and supplemental diets and organic pond fertilisation on production of *Macrobracheium rosenbergii* and associated benthic macroinvertebrate populations. Aquaculture 138, 169-180.

Wahab, M.A., Azim, M.E., Mahmud, A.A., Kohinoor, A.H.M. and Haque, M.M. 2001. Optimisation of stocking density of Thai silver barb (*Barbodes gonionotus* Bleeker) in the duckweed-fed four species polyculture system. Bangladesh J. Fish. Res. 5(1), 13-21.

Worsham, R.L., 1975. Nitrogen and phosphorus levels in water associated with a channel catfish (*Ictalurus punctatus*): Feeding operation. Trans. Am. Fish. Soc. 104, 811-815.

Zaher, M., Begum, N.N., Hoo, M.E., Begum, M. and Bhuiyan, A.K.M.A., 1995. Suitability of duckweed, *Lemna minor* as an ingredient in the feed of tilapia, *Oreochromis niloticus*. Bangladesh J. Zool. 23(1), 7-12.

Chapter Eight

Apparent digestibility coefficient of duckweed (*Lemna minor*), fresh and dry for Nile tilapia (*Oreochromis niloticus*)

Submitted to Aquaculture Research as:

Saber A. El-Shafai, Fatma A. El-Gohary, Johan A.J. Verreth, Johan W. Schrama and Huub J. Gijzen. Apparent digestibility coefficients of duckweed (*Lemna minor*), fresh and dry for Nile tilapia (*Oreochromis niloticus*).

Apparent digestibility coefficient of duckweed (*Lemna minor*), fresh and dry for Nile tilapia (*Oreochromis niloticus*)

Abstract

Dry matter (DMD), protein (PD), ash (AD), fat (FD), gross energy (ED) and phosphorus (PhD) digestibility coefficients were determined for five different iso-N fish diets fed to Nile tilapia (*Oreochromis niloticus*). Control diet contained, fish meal (35%), corn (29%), wheat (20%), wheat bran (10%), fish oil (3%), diamol (2%) and premix (1%). Partial replacement of dry matter of fishmeal, corn grain, wheat grain, wheat bran and fish oil by 20% and 40% of dry matter of duckweed, dry and fresh was performed. Diets of the treatment 1 and 2 included 20% and 40% of duckweed, respectively in a dry form. In treatment 3 and 4, tilapia received formulated diets 4 and 5 in addition to 20% and 40% fresh duckweed providing the same amount of dry matter and protein as in control. Specific growth rates of tilapia were 1.51±0.07, 1.38±0.03, 1.31±0.06, 1.44±0.02 and 1.33±0.05, in control and treatments 1 to 4. There was no significant difference between SGR for control diet and diet with 20% fresh duckweed while the other treatment groups had significantly lower SGR. All of the treatment diets provide good values for feed conversion ratios (FCR) and protein efficiency ratio (PER). The average values of FCR were 0.91, 0.98, 1.10, 0.98 and 1.10 for control and treatments 1 to 4 respectively. The average values of PER were 3.0, 2.8, 2.5, 2.9 and 2.6 in control and treatments 1 to 4, respectively. Except for treatment 2, which has higher significant faeces recovery (collected -faeces to total faeces excreted) percentage (p<0.05) there was no significant differences between control and treatments. Faeces recovery in treatment 1 was significantly lower (p<0.05) than the other treatments but not control. There was no significant difference between the dry matter content of faeces except in treatment 4, which had lower significant value in comparison to control and treatment 2. DMD ranged from 61.8% in treatment 4 to 85.2% in control. All the diets have high protein digestibility (88.4%-93.9%) and high-energy digestibility (78.1%-90.7%). Digestibility coefficients were determined for dry and fresh duckweed. DMD of duckweed were 66.8, 63.3, 45.8 and 28.3 in treatments 1 to 4, respectively. PD values were 78.4, 79.9, 77.6 and 75.9 while ED values were 59.8, 60.9, 64.5 and 58.4 in treatments 1 to 4, respectively. Body composition showed that tilapia fed diets with duckweed contain significantly (p<0.05) higher phosphorus and protein content and significantly (p<0.05) lower lipid content. Contrary, tilapia fed control diet has significant higher (p<0.05) dry matter content and lower ash content.

Keywords: duckweed digestibility, fresh duckweed, dry duckweed, Nile tilapia

1. Introduction

Fishmeal is the major source of dietary protein in fish feeds. Partial or complete replacement of fishmeal with plant protein sources could be of considerable economic advantage even if this would be accompanied by a moderate reduction in feed quality and feed utilisation efficiency. In the last decades, a considerable amount of research on the substitution of fishmeal by vegetable protein sources has been conducted

(Gomes *et al.*, 1995; McGoogan and Reigh, 1996). Leguminous seeds were identified as alternative protein source in fish feed (De Silva *et al.*, 1988; Robaina *et al.*, 1995; Eusebio and Coloso, 1998; Burel *et al.*, 2000; Eusebio and Coloso, 2000; Gouveia and Davies, 2000). In most developing countries, leguminous seeds constitute part of the human diet and its use for aquafeeds would compete with human food security. Low cost and readily available sources of vegetable protein not eaten by human have to be explored. Duckweed (*Lemnaceae*) is an aquatic plant with high protein content and is being used in waste stabilization ponds for wastewater purification. This plant could represent a good source of high quality protein for fish farmers in developing countries. Only little research work has been focused on the use of duckweed as fish feed. Digestibility measurements could be used to evaluate the nutritional value of feed ingredients and the effect of diet composition on apparent nutrients absorption (Nengas *et al.*, 1995; Sullivan and Reigh, 1995; Hasan *et al.*, 1997; Hossain *et al.*, 1997; Jones and De Silva, 1997; Krogdahl *et al.*, 1999). Beside temperature (Kim *et al.*, 1998), manufacturing process and treatment of feed (Hossain and Jauncey, 1989; Hajen *et al.*, 1993; Gomes *et al.*, 1993; Burel *et al.*, 2000; Eusebio and Coloso, 2000; Zhu *et al.*, 2001), both the type of marker and faeces collection method significantly influence the digestibility coefficients (Van Den Berg and De La Noue, 2001). Difficulties to collect total faeces output in an aquatic system make it necessary to use an indirect marker technique. The most reliable and consistent estimates of nutrients digestibility were obtained by using chromic oxide and crude fibre as dietary markers, while the performance of acid washed sand and polyethylene as dietary markers was disappointing (Tacon and Rodrigues, 1984). The values of digestibility were significantly lower, when using titanium dioxide instead of chromic oxide in rainbow trout (Weatherup and McCraken, 1998). On the other hand, Sales and Britz (2001) reported that Acid-Insoluble Ash (AIA) as an internal marker was the only marker that yielded consistent and realistic apparent digestibility coefficients. Chromic oxide and crude fibre levels in faeces were either lower or similar to their respective levels in feed, resulting in negative apparent digestibility coefficient (Sales and Britz, 2001). The naturally occurring AIA or AIA supplemented with Celite is a good alternative marker for chromic oxide and more accurate and realistic than acid-washed sand (Goddard and Mclean, 2001). For this reason, AIA was used as digestibility marker in the current research. To our knowledge, the apparent digestibility coefficients (ADC) of fresh or dry duckweed, as feed ingredient for tilapia are unknown.

In the present study, the ADC of dry matter, organic matter, ash, crude fat, protein and phosphorus in dry and fresh duckweed (*Lemna minor*), was determined for Nile tilapia (*Oreochromis niloticus*). Evaluation of duckweed as fish feed ingredient for Nile tilapia (*Oreochromis niloticus*) was subsequently assessed.

2. Materials and methods

2.1. Fish and husbandry

The experiment extended for 49 days. Juveniles Nile tilapia (*Oreochromis niloticus*) of about 90 grams were raised in the hatchery of the Fish Culture and Fisheries Group (Animal Science Department) at Wageningen University, The Netherlands. The fish were randomly distributed over 16 glass aquaria (70 litre) in-groups of 40 fish each. Control in quadruplicate and the treatments in triplicates, were randomly assigned to the 16 tanks. Tanks were linked with a thermo-regulated water re-circulating system, comprising of settling tank for suspended solids removal and an aerobic bio-filter (trickling filter) as ammonia nitrifying unit. Ammonia, nitrite and nitrate were monitored three times a week to assess the performance of the bio-filter and water quality in aquaria. The water pH was maintained within the suitable range of Nile tilapia by addition of sodium bicarbonate to the effluent of the trickling filter before feeding to the fish tanks. Water temperature, pH, dissolved oxygen (DO) and electric conductivity (EC) were monitored daily. The water flow rate in each of culture tanks was kept at 6 litres per minute. A constant photoperiod of 12 hours light/12hours day was maintained. At the beginning of the experiment 10 fish were randomly selected, weighed, killed by using an over dose of Trimethylsilyl (TMS) (8 grams dissolved in 10 litre of sodium bicarbonate solution (16 grams/litre)) for 10 minutes and then frozen at -20 °C for subsequent body composition analysis.

Table (1): Diets composition as % of individual ingredients (dry matter bases)

Ingredients	Diet 1	Diet 2	Diet 3	Diet 4	Diet 5
Fishmeal	35	27.78	20.57	34.74	34.3
Dry duckweed[1]	0	20	40	0	0
Corn	29	23.02	17.04	28.78	28.42
Wheat	20	15.88	11.75	19.89	19.6
Wheat bran	10	7.94	5.88	9.93	9.8
Fish oil	3	2.38	1.76	2.98	2.94
Diamol	2	2	2	2.5	3.33
Premix	1	1	1	1.25	1.67

[1] *Lemna minor*

Table (2): Proximate compositions of diets as, g/kg dry matter except dry matter as g/kg diet and energy as KJ/g dry matter

Parameter	Diet 1	Diet 2	Diet 3	Diet 4	Diet 5	Dry dw[*]	Fresh dw[*]
Dry matter	909 ±0.4	878 ±0.2	878±0.6	908 ±0.3	909 ±0.4	873 ±0.6	75.2±5.8
Organic matter	907 ±1.1	888 ±0.1	871 ±0.1	902 ±0.7	894 ±0.3	838 ±0.5	876 ±0.5
Ash	93±1.1	112±0.1	129±0.1	98±0.7	106±0.3	162 ±0.5	124 ±0.5
AIA[**]	18.2±0.3	18.9±0.3	19.8±0.1	23.2±0.3	30.5±0.1	3.34±0.1	0.86±0.1
Crude Protein	361±2.9	368±2.7	373±1.5	354±1.8	350 ±2.6	410 ±2.6	361 ±1.6
Crude Fat	101±0.5	86.3±0.1	80.4±0.5	102.6±1	99.7±0.1	44.4±0.2	60.1±0.5
Crude fibre	20±0.6	31.3±1.1	43.1±0.6	19.3±1.0	18.4±0.1	93.8±2.0	109±0.6
Phosphorus	11.7±0.2	12.6±0.1	13.7±0.2	11.5±0.2	11.3±0.1	16.9±0.1	9.6±0.03
Gross Energy	20.3±0.1	19.7±0.1	19.3±0.1	19.1±0.1	19.9±0.0	17.9±0.1	19.0±0.1

[*] duckweed and

[**] AIA is acid insoluble ash

2.2. Dietary Composition

Five experimental diets were formulated according to the ingredient composition, shown in Table 1. The control diet (diet 1) was formulated to meet the known requirements of tilapia. The macronutrients were separately grinded, powdered ($\leq$ 200μm particle size) and homogeneously mixed. The micro-ingredients were mixed and added solely to the macronutrients. The pellet diets were prepared on a cold feed pellet machine with a 2- mm die. The resulting pellets were air-dried, sieved, packed and stored at 5 °C. The nutrient composition of the formulated diets and tested duckweeds are presented in Table 2.

Table (3): Nutrients content of treatment diets in g/100 g of dry matter except for gross energy

Item	Control	Treatment 1	Treatment 2	Treatment 3	Treatment 4
Organic matter	90.69	88.84	87.06	89.66	88.67
Ash	9.31	11.16	12.94	10.34	11.33
AIA	1.82	1.89	1.98	1.87	1.86
Crude Protein	36.1	36.79	37.29	35.57	35.44
Crude Fat	10.1	8.63	8.04	9.41	8.39
Crude fibre	1.9	3.13	4.31	3.73	5.47
Phosphorus	1.17	1.26	1.37	1.12	1.07
Gross Energy (KJ/g)	20.29	19.73	19.34	19.07	19.53

2.3. Feeding pattern

Control group and treatment 1 and 2 received 80 gram from diet 1, 2 and 3 respectively on a daily basis. For the treatments of fresh duckweed (3 and 4), the fish received 80% (64 grams) and 60% (48 grams) from diets 4 and 5; in addition to 20% (equal to 16 gram of diet 4) and 40% (equal to 32 gram of diet 5) of fresh duckweed, respectively based on the dry matter content of the diets (4 & 5) and duckweed. The fish was fed by hand twice a day at 8:30 a.m and 1:30 p.m. The duckweed (*Leman minor*) was bought from a Dutch company where they grow duckweed in a greenhouse using artificial fertilisers. Air-dried (°C 70) duckweed was received in one package three weeks before the experiment. This dry duckweed was used in the processing of diets 2 and 3. The fresh duckweed had being received weekly during the experiment. Dry matter analysis was carried out three times for the weekly sample and the samples used for dry matter analysis stored for composition analysis by the end of the experiment. Compositions of the treatment diets and feed components applied to the tanks are presented in Table 3 and Table 4.

Table (4): Feeding levels of dietary nutrients applied to the treatment tanks in g tank^{-1} day^{-1} (except for energy, which is in KJ tank^{-1} day^{-1})

Item	Control	Treatment 1	Treatment 2	Treatment 3	Treatment 4
Dry matter	72.72	70.56	70.08	72.64	72.64
Organic matter	65.95	62.69	61.01	65.13	64.41
Ash	6.77	7.78	9.07	7.51	8.23
Crude protein	26.25	25.96	26.13	25.84	25.74
Crude fat	7.27	6.09	5.63	6.84	6.09
Crude fibre	1.38	2.21	3.02	2.71	3.97
Dietary Phosphorus	0.85	0.89	0.96	0.81	0.78
Gross Energy (KJ)	1475	1392	1355	1386	1418

2.4. Faeces collection

Weatherup and McCraken, (1998) demonstrated that both, faeces collection by stripping or by removal from the tank, have inherent weaknesses. However, the tank method gave accurate and replicable estimates of apparent digestibility of diets for fish. The objections to direct faeces sampling from the intestine (stripping) relate to the fact that the faecal material may be removed prior to completion of the natural retention time thereby reducing the digestion and absorption capacity and resulting in poor digestibility. In the current study, the mesh collector (Choubert) technique was used (Choubert et al., 1982). After a 7 days conditioning period, the faeces were collected, 6 days a week for four weeks. Eight collectors were used for 16 tanks, changing the collectors over the tanks every three days.

2.5. Analytical Methods

Water temperature, dissolved oxygen (DO), electric conductivity (EC) and pH were monitored using portable instruments while ammonia, nitrite and nitrate were measured using water quality test kits (Aquamerck). Proximate analysis of diets, duckweed, faecal samples and fish biomass samples were made following the usual procedures. Dry matter was measured after drying in an oven at 103 °C until constant weight. Ash content was determined by incineration in a muffle furnace at 550 °C. The crude protein was measured by the Kjeldahl method (N×6.25) after acid digestion. Lipid was extracted by petroleum ether in a soxhlet apparatus. Crude fibre was measured by the acid/base digestion method. Phosphorus was analysed by the vanado-molybdate method after combustion at 550 °C and digestion with acid. AIA was determined by based-acid digestion method. Energy content was measured by direct combustion in an adiabitic bomb calorimeter.

2.6. Growth performance

In the current experiment the growth period of tilapia was 49 days. At the end of the trial, the fish biomass in each tank was weighed in a group and the number of fish was counted; the average individual fish weight was determined and specific growth rates (SGR) were calculated per tank using the following expression:

$$SGR = \frac{(\ln Wf - \ln Wi)}{days} \times 100 \qquad (1)$$

Where, Wi and Wf are initial and final mean body weight respectively.

Total feed intake was recorded in each treatment and the feed conversion ratios (FCR) and protein efficiency ratio (PER) were calculated at the end of the trial based on the following equations:

$$FCR = \frac{total\ feed\ ingested\ (dry\ weight)}{fish\ weight\ gain\ (wet\ weight)} \qquad (2)$$

$$PER = \frac{fish\ weight\ gain\ (wet\ weight)\ (g)}{total\ protein\ ingested\ (g)} \qquad (3)$$

2.7. Apparent digestibility measurement

The Apparent dry matter digestibility (ADMD) in the diets was calculated using the following formula:

$$\text{ADMD in diets (\%)} = 100 \times \left[1 - \frac{\% \text{ MD}}{\% \text{ MF}} \right] \qquad (4)$$

Where, MD is the marker in diet and MF is the marker in faeces.

The apparent nutrients digestibility (AND) in the diets was calculated using the following formula:

$$\text{AND in diets (\%)} = 100 \times \left[1 - \frac{[\% \text{ MD}][\% \text{ NF}]}{[\% \text{ MF}][\% \text{ ND}]} \right] \qquad (5)$$

Where, NF is the nutrient in faeces and ND is the nutrient in the diet.

The apparent dry matter digestibility (ADMD) in the test ingredient was calculated as follows:

$$\text{ADMD in TI (\%)} = \frac{\left[\text{ADC}_{TD} - [\% \text{ RD}][\text{ADC}_{RD}] \right]}{\% \text{ of TI}} \qquad (6)$$

Where ADC_{TD} is the apparent digestibility coefficient of the test diet, TD is the test diet, ADC_{RD} is the apparent digestibility coefficient of the reference diet, RD is the reference diet and TI is the test ingredient.

The ADC of organic matter, ash, fat and nutrient (%) was calculated using the formula applied by Sugiura *et al.* (1998):

$$\text{ADC (\%)} = \frac{[\text{TD}_{nutrient} \times \text{ADC of TD}_{nutrient}] - [\% \text{ RD} \times \text{ADC of RD}_{nutrient} \times \text{RD}_{nutrient}]}{[\% \text{ TI} \times \text{TI}_{nutrient}]}$$

$$(7)$$

Where $\text{TD}_{nutrient}$ is the nutrient concentration in the test diet, $\text{RD}_{nutrient}$ is the nutrient concentration in the reference diet and $\text{TI}_{nutrient}$ is the nutrient concentration in the test ingredient.

$$\% \text{ Faeces recovery} = \frac{[\% \text{ MF} \times \text{QF}]}{[\% \text{ MD} \times \text{QD}]} \qquad (8)$$

Where MF is the marker in faeces, QF is the quantity of faeces collected, MD is the marker in diet and QD is the quantity of diet given.

2.8. Data Analysis

Data are presented as means with standard deviation. To test the statistical differences between the treatment groups including control, data were subjected to a one-way analysis of variance (ANOVA). 95% probability was chosen to test the significance of differences between the treatments.

3. Results

3.1. Water quality

The water quality parameters at the inlet and outlet of the re-circulation unit are presented in Table 5. All of the parameters lie within the desirable limits for Nile tilapia.

Table 5: Water quality parameters in the inlet and outlet of the re-circulating system

Water sample	Temperature (°C)	pH	DO (mg O_2 l^{-1})	EC (μ mhos cm^{-1})	Ammonia (mg N l^{-1})	Nitrite (mg N l^{-1})	Nitrate (mg N l^{-1})
Inlet	27.4-29	6.9-8.2	7.7±0.4	227±29	Not detected	0.02±0.02	7.1±4.4
Outlet	27.2-29	6.5-7.6	6±0.4	226±28	0.07±0.09	0.06±0.02	7.1±4.4

3.2. Growth performance

Individual feeding levels of nutrients applied to each tank were calculated and presented in table 4. In treatment 3 and 4, despite the presence of two different feed types, i.e. feed pellets and fresh duckweed, the fish fed eagerly on both diets and consumed all of the duckweed. The fresh duckweed in these treatments was completely consumed.

Table 6: Growth performance of tilapia (*Oreochromis niloticus*) in the experimental tanks

Item	Control	Treatment 1	Treatment 2	Treatment 3	Treatment 4
Initial body weight (g fish^{-1})	89.3±2.1	91.5±0.9	88.5±1.5	88.7±0.3	87.5±1.9
Final body weight (g fish^{-1})	186.9±2.1	180.1±0.7	168.1±5	179.4±1.4	168.1±2.7
Stoking density (fish tank^{-1})	40	40	40	40	40
Total weight gain (g fish^{-1})	97.6±4.2	88.6±1.6	79.6±4.9	90.7±1.6	80.6±2.8
Total feed intake (g fish^{-1})[**]	89	86.4	86	89	89
Total protein intake (g fish^{-1})	32.16	31.8	32.01	31.65	31.53
Specific growth rate (SGR)	1.51±0.07[a]	1.38±0.03[b]	1.31±0.06[b]	1.44±0.02[a]	1.33±0.05[b]
Feed conversion ratio (FCR)[*]	0.91±0.04[a]	0.98±0.02[ab]	1.08±0.07[bc]	0.98±0.02[ab]	1.11±0.04[c]
Protein efficiency ratio (PER)	3.04±0.13[a]	2.78±0.05[b]	2.49±0.15[c]	2.86±0.05[ab]	2.56±0.09[c]
% of mortality	1.9	0.8	0.8	0	0.8

[*] g dry feed/g fish weight gain

[**] g dry feed/fish

Table (6) shows the growth performance of juvenile tilapia (*Oreochromis niloticus*) fed control and treatment diets. Fish receiving the control diet had significant higher specific growth rates than those receiving diets containing 20% (T1) and 40% (T2) dry duckweed as well as those receiving diets containing 40% fresh duckweed (T4). There was no significant difference between the SGR of control tilapia and those fed the diet containing 20% (T3) fresh duckweed. In spite of the better significant FCR in control tilapia, all the treatments showed good values (especially in case of 20% duckweed, both for dry or fresh). The control group had significant higher PER than the treatment groups T1, T2, and T4 but no significant difference was observed with T3. Despite the higher SGR, FCR and PER of the control, all the treatment groups showed good values in comparison to literature values.

Table 7: Apparent digestibility coefficients (%) of the treatment diets in the experimental tanks

Parameter	Control	Treatment1	Treatment2	Treatment3	Treatment4
Apparent dry matter digestibility (ADMD)	85.2 ± 0.2^a	77.3 ± 2.2^b	71.7 ± 0.8^c	76.4 ± 1.2^b	61.8 ± 0.3^d
Apparent organic matter digestibility (AOMD)	87.5 ± 0.2^a	79.3 ± 2.2^b	73.6 ± 0.8^c	79 ± 1.2^b	64.9 ± 0.3^d
Apparent ash digestibility (AAD)	62.9 ± 1.5^a	61.4 ± 2.1^{ab}	58.8 ± 1.1^b	53.9 ± 1^c	38 ± 1.1^d
Apparent protein digestibility (APD)	93.9 ± 0.1^a	91.2 ± 0.5^b	89.7 ± 0.4^c	92 ± 0.3^b	88.4 ± 0.1^d
Apparent fat digestibility (AFD)	96.5 ± 1.1^a	93.9 ± 0.5^b	90.9 ± 1^d	92 ± 0.3^c	82.3 ± 0.8^e
Apparent phosphorus digestibility (APhD)	86.7 ± 0.9^a	84.8 ± 1.2^{ab}	83.8 ± 0.9^b	85.2 ± 0.9^a	81.1 ± 1.3^c
Apparent energy digestibility (AED)	90.7 ± 0.3^a	84.8 ± 1.7^b	79.3 ± 1.5^c	85.5 ± 1.5^b	78.1 ± 3.7^c

P<0.05 (Statistical analysis between values within the same row and values with different superscript (a, b, c and d) are statistically significant)

3.3. *Apparent digestibility coefficients of the treatment diets and duckweed*

The mean apparent digestibility coefficients for the dry matter and nutrients of treatment diets are shown in Table 7. Protein, which is the major growth-limiting factor in the fish diets, has an apparent digestibility coefficient of more than 90% in the control diet and in treatment diets 1 and 3 (20% duckweed addition) while the value was slightly lower in treatment 2 and 4 (40% duckweed addition). The average values of energy digestibility were 90.7% in control diet and in the range of 78.1%-85.5% in the treatment diets.

The apparent digestibility coefficients of the nutrients in dry and fresh duckweed are presented in Table 8. Apparent dry matter digestibility coefficients of dry and fresh duckweed in case of 20% are not significantly different while the organic matter digestibility was significantly higher in the dry duckweed. However, 40% dry duckweed showed higher values compared to the fresh duckweed except the AED where there was no significant difference. In case of dry duckweed there is no significant difference between 20% and 40% except for fat digestibility, which is significantly lower in T2. In case of fresh duckweed the 20% had significantly higher values for dry matter, organic matter and crude fat while values of protein, phosphorus and energy digestibility are statistically similar.

Table 8: Apparent digestibility coefficients (%) of duckweed (*Lemna minor*)

Parameter	Treatment 1	Treatment 2	Treatment 3	Treatment 4
Apparent dry matter digestibility (ADMD)	51.83±4.4[a]	51.4±1.9[a]	44.7±0.2[a]	26.8±0.9[b]
Apparent organic matter digestibility (AOMD)	66.83±7.4[a]	63.3±2.3[a]	45.8±0.6[b]	28.3±0.7[c]
Apparent protein digestibility (APD)	78.4±2.2[a]	79.0±0.9[a]	77.6±1.5[ab]	75.9±0.3[b]
Apparent fat digestibility (AFD)	73.6±4.0[a]	60.1±4.9[b]	71.3±2.3[a]	43.8±2.7[c]
Apparent phosphorus digestibility (APhD)	75.9±4.6[a]	79.8±1.8[a]	74.6±5.0[ab]	67.4±3.6[b]
Apparent energy digestibility (AED)	59.8±9.0[a]	60.9±3.8[a]	64.5±7.7[a]	58.4±9.5[a]

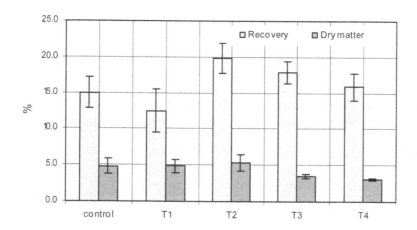

Fig. 1: Percentage of faeces recovery and its dry matter content

3.4. Faeces recovery and nitrogen mass balance

As presented in Figure 1, the %faeces recovery (collected faeces from the total excreted faeces) is respectively, 15, 12.5, 19.8, 17.9 and 15.9% in control and treatments 1 to 4. There is no significant difference between control and treatments except for T2, which had a higher value than the control. Faeces recovery in T1 was significantly lower than in the other treatments. For the dry matter content of faeces there was no significant difference except for T4, which had a lower value in comparison to control and T2. Nitrogen mass balance in control and treatment groups are presented in Figures 2. The range of nitrogen recovery (retention in fish biomass) percentages in control and treatment groups was 40.6-47.5% while soluble excreta and particulate nitrogen in faeces were 45.4-47.7 and 6.1-11.6, respectively.

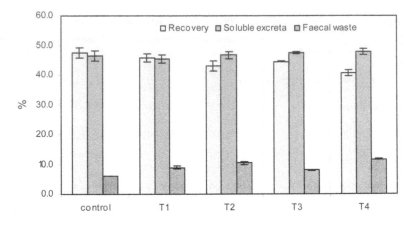

Fig. 2: Nitrogen mass balance in control and treatment groups

3.5. Body composition

The body compositions of tilapia at the start and at the end of the experiment are presented in Table 9. The results showed that tilapia fed control diet had a lower protein and phosphorus content (g per 100 gram of dry body matter) while dry matter and crude fat were significantly higher than in tilapia receiving diets with duckweed supplements. Tilapia fed treatment diets had significantly higher ash content. The fat and protein gain, as gram/fish were significantly high in tilapia fed control diets.

Table 9: Body composition[*] of tilapia in control and treatment tanks

Parameter	Starting	Control	Treatment 1	Treatment 2	Treatment 3	Treatment 4
Dry matter	30±0.15	30.7±0.5[a]	30.4±0.6[ab]	29.1±0.7b[c]	29.2±0.3[c]	28.7±0.8[c]
Ash	13.9±0.32	14.1±0.7[a]	15.9±0.2[b]	15.8±0.6[b]	15.7±0.7[b]	16.1±0.2[b]
Crude protein	50±0.12	50±1.2[a]	51.8±1[ac]	55.5±0.7[b]	52.4±0.4[c]	53.9±1.6[bc]
Protein gain[1]	-	15.3±0.6[a]	14.6±0.5[ab]	13.8±0.5[bc]	14.15±0.0[b]	12.8±0.3[c]
Phosphorus	2.5±0.14	2.4±0.1[a]	2.7±0.1[b]	2.7±0.2[b]	2.7±0.1[b]	2.8±0.1[b]
Crude fat	31.4±0.35	33.6±0.7[a]	29.9±0.6[b]	26.7±0.9[c]	29.3±1.6[bc]	26.7±1.6[c]
Fat gain[1]	-	10.9±0.7[a]	7.8±0.8[b]	4.7±0.4[c]	7.0±0.9[b]	4.7±1.0[c]
Gross Energy	21.0±0.2	20.5±0.6[a]	20.3±1.1[ab]	18.5±0.8[bc]	18.9±1.1[ab]	17.5±0.7[bc]

[*] As % of dry matter except dry matter as % of fresh weight and energy as KJ/g of dry matter

[1] as gram protein or fat per fish

4. Discussion

4.1. Growth performance

In this experiment all of the fresh duckweed were well accepted and consumed in the presence of feed pellets. Contrary to our results, Hajen et al., (1993) and Hasan et al., (1997) reported a low intake of feed containing increasing levels of vegetable protein. The ranges of SGR, FCR and PER in this study were 1.3-1.5, 0.9-1.1 and 2. 5-3.0, respectively. These results are better than the reported ranges of SGR (0.9-1.4) and FCR (1.6-2.1) of tilapia with weights ranging from 15-68 g/fish and receiving fishmeal-based diets (El-Sayed, 1992; Papoutsoglau and Tziha, 1996; Mbahinzireki et al., 2001; Middleton et al., 2001). This might be attributed to the good nutritional quality of the combined feeding of fishmeal-based diets with duckweed as used in the present experiment in comparison to the combination of fishmeal-based diets with other plant ingredients as used by other authors. Improved feeding techniques combined with improvements in the quality of feed have lowered FCRs in present-day freshwater fish farming with FCR values close to 1.0 (Enell, 1995; Iversen, 1995). This is similar to our data. The slight reduction in the growth performance of tilapia in the duckweed-supplemented groups in comparison to the control could be attributed to the lower protein and energy intake in these treatment groups (Table 4). Weight gain and FCR are negatively affected as the dietary protein and energy decrease (Lee and Kim, 2001)). The lower digestibility in the treatment diets is also a reason. Deficiency of duckweed in the amino acids cysteine and methionine (Hammouda et al., 1995) could also be a reason. Hossain and Jauncey (1989) reported that the deficiency of linseed in some essential amino acids, namely lysine, methionine and therionine, resulted in a reduction of growth performance of carp fed diets with 25% linseed. Also other authors (Tantikitta and Chimsung, 2001; Yamamoto et al., 2001; Lovel, 1991; El-Dahhar and El-Shazly, 1993) reported on the role of essential amino acids lysine, methionine and cystine in enhancing the growth performance of fish.

4.2. Apparent digestibility coefficients of the treatment diets

In this trial good values were obtained in all diets for the apparent digestibility coefficients. The dry matter digestibility of the control diet (85.2) differed significantly from the one in the test diets, but the values in treatments 1, 2, and 3 are still considered to be good within the range of 71.7-77.3. This range is slightly higher than the range (64%-71%) of DMD that has been reported for fishmeal-based diet with soybean, wheat bran, wheat flour, cornstarch, corn oil, Cod liver oil and vitamin and mineral premix, fed to Oreochromis aureus (Goddard and McLean, 2001). In all treatment diets, ADC of protein was in the range of 88.4%-92%, which is better than the range of 78%-90% reported for common carp fishmeal-based diets, which included 25% and 50% of linseed, groundnut or sesame (Hasan et al., 1997). The higher values in our study could be attributed to the lower percentages of plant ingredients in the dietary treatments. Significantly lower values of protein digestibility, fat digestibility and energy digestibility were obtained in the treatment diets in comparison to the control. This could be attributed to the higher levels of ash and crude fibre in the treatments with duckweed addition. Chaimongkol and Boonyarat (2001) reported negative effects of dietary ash on protein and phosphorus digestibility. In turbot, low digestibility coefficients were attributed to high fibre content especially to acid detergent and neutral detergent fibres in rapeseed meal (Burel *et al.*, 2000). In a number of cases (Brown *et al.*, 1989; Kim *et al.*, 1998) plant-based proteins are reported to be more digestible than those of animal origin, whereas other authors (Hossain and Jauncey, 1989; Sullivan and Reigh, 1995; McGoogan and Reigh, 1996) reported the opposite. The partial replacement of brown fishmeal by vegetable protein up to 66%, without negative effects, was demonstrated in rainbow trout (Gomes *et al.*, 1995). Fagbenro (2001) observed no significant difference between the protein digestibility coefficient from animal and plant origin sources for *Heterotis niloticus*. In this study duckweed (plant protein) digestibility was lower than fishmeal digestibility, which shows that the comparison between digestibility of plant and animal-based diets is strongly depending on the source and quality of both animal and plant protein.

There was no significant difference between the PD of test diets in T1 and T3 with average values of 91.2 and 92.0 % while a significantly higher value was obtained in T2 (89.7) in comparison to T4 (88.4). This might be attributed to the high moisture content and volume of fresh duckweed, which causes dilution of the enzyme concentration. Also, the processed dry duckweed had smaller particle size, which resulted in a larger specific surface area of the substrate.

The lower values of phosphorus digestibility in treatment diets in comparison to the control could be attribute to the presence of high amounts of plant tissue in the form of duckweed. Phosphorus is mainly present conjugated in phytin in plant tissue (Weerasinghe *et al.*, 2001) and this is the reason why phytase treated diets have higher phosphorus digestibility (Papatryphon and Soares Jr, 2001; Papatryphon *et al.*, 1999; Oliva-Teles *et al.*, 1998). Lower phosphorus digestibility of treatment diets may also have resulted from plant protein in the diets since some products from plant protein hydrolysis have been reported to cause decrease in the phosphorus solubility by altering the ion distribution (Weerasinghe *et al.*, 2001).

The higher significant values of the apparent digestibility coefficients of crude fat and gross energy in control could be attributed to the higher amounts of crude fibre and ash in the treatment diets.

4.3. Apparent digestibility coefficients of duckweed

Since feed formulation should be based on nutrient availability, reliable data on the digestibility of different ingredients are important. Such data are available for some commonly used ingredients. In this experiment, apparent digestibility coefficients of duckweed protein in all the treatment diets, lie within the range of 75.9%-79%, which is comparable to the range of 75.4%-83.8% reported for cottonseed meal (Reigh *et al.*, 1990; Nengas *et al.*, 1995; Sullivan and Reigh, 1995; Fagbenro, 1996; Fagbenro, 2001). The slight higher maximum value in cottonseed meal might be attributed to the high oil content in comparison to duckweed since it has been reported that increment in lipid content of diet or ingredient is accompanied with increase in protein digestibility (Gomes *et al.*, 1993; Hajen *et al.*, 1993; McGoogan and Reigh, 1996; Martinez-Palacois, 1998). Our data showed lower APD than the reported values (80%-94.8%) for soybean meal (Reigh *et al.*, 1990; Hossain *et al.*, 1992; Nengas *et al.*, 1995; Sullivan and Reigh, 1995). Carbohydrates in duckweed consist to a large extend of fibres instead of starch, which is a major carbohydrate in grains like soybean. Fish appears to digest starchy plant materials more effectively than roughage plant meal (Reigh *et al.*, 1990). Soybean contains more fat and protein than duckweed. These two parameters positively influence dry matter and protein digestibility while crude fibre and ash have negative effects (Gomes *et al.*, 1993; Hajen *et al.*, 1993; McGoogan and Reigh, 1996; Martinez-Palacios, 1998). The APD values were slightly lower than the reported range (83%-96%) for extruded peas, extruded lupine and rapeseed in the diets of turbot and trout (Burel *et al.*, 2000). This might be attributed to the effect of processing heat in case of the extruded pellets. It has been reported that the inhibitory effect of anti-nutritional factors in plant ingredients could be removed or reduced by heat treatment (Gomes *et al.*, 1993; Burel *et al.*, 2000). The dry matter digestibility of duckweed in this experiment even in case of 40% fresh duckweed is comparable to the reported range of 27.5%-50% for wheat middling, wheat flour, corn grain and yellow maize (Sullivan and Reigh, 1995; Fagbenro, 1996). The non-significant difference of the AED between the 40% fresh duckweed treatment and dry duckweed treatments despite the low digestibility of dry matter, organic and fat is questionable. This might be due to the non-complete mixing of the duckweed with the feed pellets containing the marker and leaching in fresh duckweed treatments. The differences between the batch and age of the fresh and dry duckweed might be also a reason. The drying process in dry duckweed might also have negative effects on the energy digestibility of the dry duckweed.

The Phosphorus digestibility in duckweed ranged between 67.4%-79.8%, which is better than most of the reported values in other plant ingredients. The AD of phosphorus in corn gluten, soybean meal and isolated soy protein was in the range of 48-59% while wheat middling was 10% for striped bass, *Morone saxatilis* (Papatryphon and Soares Jr, 2001). Bai *et al.* (2001) obtained a phosphorus digestibility of 22% with rock fish (*Sebastes schlegeli*) while Weerasinghe *et al.* (2001) reported a phosphorus digestibility of 8.5%, 22%, 47%, 55% and 75% for corn gluten, soybean meal, wheat flour, wheat middling and wheat gluten in the diet of trout fish.

Few experiments dealt with digestibility of aquatic plants and wild grasses. The total dry matter digestibility of fresh aquatic macrophyte, *Hydrilla verticella* by *Etroplus suratensis* varied between 35-52% while the values for protein and fat were 59-71% and 65-70%, respectively (De Silva and Perera, 1983). Hydrilla is a submerged aquatic plant that is characterised by low lignin and cellulose fibre, and a muco-polysaccharide layer, which might be the reason why the dry matter digestibility of Hydrilla is slightly higher than in duckweed. The low protein digestibility in Hydrilla could be attributed to the presence of some protein in conjugation with carbohydrates in the muco-polysaccharide molecule. The presence of carbohydrate moiety provides a barrier reducing the attack of protease enzymes with the protein. Our results are better than those by Buddington, (1979) who reported values of 29% ADMD, 75% APD and 76% AFD of fresh *Najas guadalupensis* grass fed to *Tilapia zilli*. This might be attributed to low fibre content of duckweed in comparison to *Najas guadalupensis* grass, which is characterized by high fibre levels. Estimation of dry matter digestibility of the duckweed species, *Lemna gibba* and *Lemna minor* for grass carp was 65% while APD was 80% (Van Dyke and Sutton, 1977). Grass carp is characterized as herbivore and appears to be able to utilise plant fibre better than omnivores (like tilapia).

4.4. Faeces recovery and nitrogen mass balance

The %faeces recovery was significantly higher in the treatment groups. This could be attributed to the higher fibre content in duckweed, which improves the consistency of faeces. The high moisture content of faeces in diets with fresh duckweed is attributed to the moisture of duckweed. The nitrogen recovery in the control diet (47.5%) was significantly higher (p<0.05) than in the treatment diets while soluble nitrogen and faecal nitrogen were lower than in the treatment diets and represented 46.4% and 6.1%, respectively. This is attributed to two main reasons. Firstly the low protein digestibility in the treatment diets in comparison to the control diet decrease the available nitrogen. Secondly, the diets containing more plant proteins as substitutes of fishmeal show higher ammonia excretion (Gomes *et al.*, 1993; Robaina *et al.*, 1995). This higher ammonia excretion was attributed to the excess of arginine and lysine present in the plant protein (Gomes *et al.*, 1993). In the treatment diets the nitrogen recovery percentage was 40.6%-45.8% while ammonia excretion and faecal nitrogen were 45.4%-47.7% and 8.0%-11.6%, respectively. These results compare well with the range of nitrogen retention (30%-58%) in Nile tilapia fed balanced diets (Abdelghany, 1998) with 35% protein in the dry matter and variable amounts of ascorbic acid from different sources. The range of ammonia excretion in this trial compares to the reported range (41-48%) for fish received diets with 15% and 45% vegetable protein (Gomes *et al.*, 1993) as well as the nitrogen recovery (43-49%) in sea bass (*Lates calcarifer*) reported by Chaimongkol and Boonyarat (2001). The wider range reported by Gomes *et al.* (1993) could be attributed to the wider range of vegetable protein ratio of the diets since he used 15% and 45% in comparison to 20% and 40% in the current trial.

4.5. Body composition

The higher moisture content and low lipid content of tilapia fed duckweed-supplemented diets might be attributed to the higher plant protein in the diets derived from duckweed. Hossain and Jauncey (1989) reported that common carp fed diets with higher level of plant protein had higher body moisture content and lower lipid content (Hossain and Jauncey, 1989). It might be attributed to the low energy content and digestibility in the treatment diets since moisture content of fish receiving high-energy diets was significantly lower than those fed low-energy diets (Lee and Kim, 2001). High protein content of tilapia fed treatment diets could be attributed to the high plant protein derived from duckweed. Robaina *et al.* (1995) demonstrated that increment in the level of vegetable protein in the diet of gilthead seabream was associated with higher protein content in the fish.

In general the lipid content of tilapia in the control and treatment diets was in the range of 26.7-33.6% of the dry matter which is higher than the maximum reported value (27%) reported by El-Sayed (2003). The low lipid content of tilapia fed duckweed-supplemented diets could be explained by 3 reasons. Vitamin c that might be present in appreciable amounts in duckweed induces lipolysis and decrease tissue lipid in tilapia by increasing its dietary concentration (Abdelghany, 1998). The higher lipid content of the control diet might be a reason that tilapia fed on control diet had higher lipid content since Giri et al. (2000) reported that the body lipid content increases with the dietary lipid increment. The higher phosphorus content of tilapia in treatment groups could also be a reason since the increment of dietary available phosphorus appears to reduce muscle and visceral fat of fish (Eya and Lovell, 1997).

5. Conclusions

1- Digestibility coefficients of duckweed (*Lemna minor*) are better than some conventional plant ingredients used in fish feed. Digestibility coefficients of diets containing 20% and 40% dry duckweed and diets with 20% fresh duckweed encourage use of duckweed in fish feed manufacturing. Use of duckweed in dry or fresh form could replace more costly and often scarcely available fishmeal-based diets.

2- Digestibility coefficients of diets containing 40% fresh duckweed are lower but economically still attractive for use in tilapia aquaculture.

3- Digestibility coefficients of dry duckweed were better in comparison to fresh duckweed especially in diets with 40% inclusion. This might be attributed to the positive effects of milling process and the bulky volume of fresh duckweed in case of 40%.

4- Use of duckweed as a replacement for fishmeal in fish diets increases the protein and phosphorus content of dry matter of fish, which might be add to product quality.

Acknowledgements

This research was supported by a grant within framework of "Wasteval" project supported by the Dutch government. The deskwork of this study was partially supported by the World Laboratory Scholarship (T-1 project). The authors wish to thank Menno ter Veld, Roland Booms and Tino Leffering of the Fish Culture and Fisheries group of Wageningen University for their help during the lab analysis of this work.

References

Abdelghany, A.E. 1998. Feed efficiency, nutrient retention and body composition of Nile tilapia, *Oreochromis niloticus* L., fed diets containing L-ascorbic acid, L-ascorbyl-2-sulphate or L-ascorbyl-2-polyphosphate. Aquacult. Res. 29, 503-510

Bai, S.C., Chai, S.M., Kim, K.W. and Wang, X.J. 2001. Apparent protein and phosphorus digestibility of five different dietary protein sources in Korean rockfish, *Sebastes schlegeli* (Hilgendorf). Aquacult. Res. 32, 99-105

Brown, P.B., Robinson, E.H., Clark, A.E. and Lawrence, A.L. 1989. Apparent digestible energy coefficients and associative effects in practical diets for red swamp crayfish. J. World Aquaculture Society 20(3), 122-125

Buddington, R.K. 1979. Digestion of an aquatic macrophyte by *Tilapia zillii* (Gervais) J. Fish Biol. 15, 449-455

Burel, C., Boujard, T., Tulli, F., Kaushik, S.J. 2000. Digestibility of extruded peas, extruded lupin, and rapeseed meal in rainbow trout (*Oncorhynchus mykiss*) and turbot (*Psetta maxima*). Aquaculture 188, 285-298

Chaimongkol, A. and Boonyaratpalin, M. 2001. Effects of ash and onorganic phosphorus in diets on growth and mineral composition of seabass *Lates calcarifer* (Bloch). Aquacult. Res. 32, 53-59

Choubert, P.; De La Noue, J. and Luquet, P. 1982. Digestibility in fish: improved device for the automatic collection of faeces. Aquaculture 29, 185-189.

De Silva, S.S. and Perera, M.K. 1983. Digestibility of an aquatic macrophyte by the Cichlid Etroplus suratensis (Bloch) with observations on the relative merits of three indigenous components as markers and daily changes in protein digestibility. J. fish. Biol. 23, 675-684

De Silva, S.S., Keembiyahetty, C.N. and Gunasekera, R.M. 1988. Plant ingredient substitutes in *Oreochromis niloticus* (L.) diets: Ingredient digestibility and effect of dietary protein content on digestibility. J Aquacult. Trop. 3, 127-138

El-Dahhar, A.A. and El-Shazly, K. 1993. Effects of essential amino acids (metionine and lysine) and treated oil in fish diet on growth performance and feed utilization of Nile tilapia, *Tilapia niloticus* (L.). Aquacult. Fish. Manage. 24, 731-739

El-Sayed, A.F.M. 1992. Effects of substituting fish meal with *Azolla pinnata* in practical diets for fingerling and adult Nile tilapia, *Oreochromis niloticus* L. Aquacult. Fish. Manage. 23, 167-173

El-Sayed, A.M. 2003. Effects of fermentation methods on the nutritive value of water hyacinth for Nile tilapia *Oreochromis niloticus* (L.) fingerlings. Aquaculture 218, 471-478

Enell, M. 1995. Environmental impact of nutrients from Nordic fish farming. Water Sci. Technol. 31(10), 61-71

Eusebio, P.S. and Coloso, R.M. 1998. Evaluation of leguminous seed meals and leaf meals as plant protein sources in diets for juvenile *Penaeus indicus*. The Israeli Journal of Aquaculture (Bamidgeh) 50(2), 47-54

Eusebio, P.S. and Coloso, R.M. 2000. Nutritional evaluation of various plant protein sources in diets for Asian sea bass *Lates calcarifer*. J. Appl. Ichthyol. 16, 56-60

Eya, J.C. and Lovell, R.T. 1997. Available phosphorus requirements of food-size channel catfish (*Ictalurus punctatus*) fed practical diets in ponds. Aquaculture 154, 283-291

Fagbenro, O.A. 1996. Apparent digestibility of crude protein and gross energy in some plant and animal-based feedstuffs by *Clarias isheriensis* (Siluriformes: Clariidae) (Sydenham 1980). J. Appl. Ichthyol. 12, 67-68

Fagbenro, O.A. 2001. Apparent digestibility of crude protein and gross energy in some plant and animal-based feedstuffs by *Heterotis niloticus* (Dupeiformes: Osteoglossidae) (Luvier 1829). J. Aquacult. Trop. 16(3), 277-282

Giri, S.S., Sahoo, S.K., Sahu, A.K. and Mukhapadhyay, P.K. 2000. Nutrient digestibility an intestinal enzyme activity of *Clarias batrachus* (Linn) juveniles fed on dried fish and chicken viscera incorporated diets. Bioresour. Technol. 71, 97-101

Goddard, J.S. and Mclean, E. 2001. Acid insoluble ash as an inert reference material for digestibility studies in tilapia, *Oreochromis aureus*. Aquaculture 194, 93-98

Gomes, E.F., Corraz, G. and Kaushik, S. 1993. Effects of dietary incorporation of co-extruded plant protein (rapeseed and peas) on growth, nutrient utilization and muscle fatty acid composition of rainbow trout (*Oncorhynchus mykiss*). Aquaculture 113, 339-353

Gomes, E.F., Rema, P. and Kaushik, S.J. 1995. Replacement of fishmeal by plant proteins in the diet of rainbow trout (*Oncorhynchus mykiss*): digestibility and growth performance. Aquaculture 130, 177-186

Gouveia, A. and Davies, S.J. 2000. Inclusion of an extruded dehulled pea seed meal in diets for juveniel European sea bass (*Dicentrarchus labrax*). Aquaculture 182, 183-193

Hajen W.E., Higgs, D.A., Beames, R.M. and Dosanjh, B.S. 1993. Digestibility of various feed stuffs by post-juvenile chinook salmon (*Oncorhynhus tshawytscha*) in seawater: 2-Measurment of digestibility. Aquaculture 112, 333-348

Hammouda, O., Gaber, A. and Abdel-Hameed, M.S. 1995. Assessment of the effectiveness of treatment of wastewater-contaminated aquatic systems with *Lemna gibba*. Enz. and Micro. Technol. 17, 317-323

Hasan, M.R., MaCintosh, D.J. and Jauncey, K. 1997. Evaluation of some plant ingredients as dietary protein sources for common carp (*Cyprinus carpio* L.) fry. Aquaculture 151, 55-70

Hossain, M.A. and Jauncey, K. 1989. Nutritional evaluation of some Bangladeshi oil seed meals as partial substitutes for fishmeal in the diet of common carp, *Cyprinus carpio* L. Aquacult. Fish. Manage. 20, 255-268

Hossain, M.A., Nahar, N., Kamal, M. and Islam, M.N. 1992. Nutrient digestibility coefficients of some plant and animal proteins for Tilapia (*Oreochromis niloticus*). J. Aquacult. Trop. 7, 257-266

Hossain, M.A., Nahar, N. and Kamal, M. 1997. Nutrient digestibility coefficients of some plant and animal proteins for rohu (*Labeo rohita*). Aquaculture 151, 37-45

Iversen, T.M. 1995. Fish farming in Denmark: Environmental impact of regulative legislation. Water Sci. Technol. 31(10), 73-84

Jones, P.L. and De Silva, S.S. 1997. Apparent nutrient digestibility of formulated diets by the Australian freshwater crayfish *Cherax destructor* Dark (Decapoda, Parastacidae). Aquacult. Res. 28, 881-891

Kim, J.D., Breque, J. and Kaushik, S.J. 1998. Apparent digestibilities of feed components from fishmeal or plant protein based diets in common carp as affected by water temperature. Aquat. Living Resour. 11(4), 269-272

Krogdahl, A., Nordrum, S., Sørensen, M., Brudeseth, L. and Røsjø, C. 1999. Effects of diet composition on apparent nutrient absorption along the intestinal tract and of subsequent fasting on mucosal disaccharidase activities and plasma nutrient concentration in Atlantic salmon *Salmo salar* L. Aquacult. Nutr. 5, 121-133

Lee, S.H. and Kim, K.D. 2001. Effects of dietary protein and energy levels on the growth, protein utilization and body composition of juvenile masu salmon (*Oncorhynchus masou* Brevoort). Aquacult. Res. 32, 39-45.

Lovel, R.T. 1991. Nutrition of aquaculture species. J. Anim. Sci. 69, 4193-4200

Martiniez-Palacios, C.A. 1988. Digestibility studies in juveniles of the Mexican cichlid, *Cichlasoma urophthalmus* (Gunther). Aquacult. Fish. Manage. 19, 347-354

Mbahinzireki, G.B., Dabrowski, K., Lee, K.J., El-Saidy, D. and Wisner, E.R. 2001. Growth, feed utilisation and body composition of tilapia (*Oreochromis sp.*) fed with cottonseed meal-based diets in a recirculation system. Aquacult. Nutr. 7, 189-200

McGoogan, B.B. and Reigh, R.C. 1996. Apparent digestibility of selected ingredients in red drum (*Sciaenops ocellatus*) diets. Aquaculture 141, 233-244

Middleton, T.F., Fekert, P.R., Boyd, L.C., Daniels, H.V. and Gallagher, M.L. 2001. An evaluation of co-extruded poultry silage and culled jewel sweet potatoes as a feed ingredient for hybrid tilapia (*Oreochromis niloticus* × *Oreochromis mossambicus*). Aquaculture 198, 269-280

Nengas, I., Alexis, M.N., Davies, S.J. and petrichakis, G. 1995. Investigation to determine digestibility coefficients of various raw materials in diets for gilthead sea bream, *Sparus auratus* L. Aquacult. Res. 26, 185-194

Oliva-Teles, A., Pereira, J.P., Gouveia, A. and Gomes, E. 1998. Utilization of diets supplemented with microbial phytase by seabass (*Dicentrachus labrax*) juveniles. Aquat. Living Resour. 11(4), 255-259

Papatryphon, E., Howell, R.A. and Soares, J.H. 1999. Growth and mineral absorption by stripped bass *Morona saxatilis* fed a plant feed stuffs based diet supplemented with phytase. J. World Aquaculture Society 30(2), 161-173

Papatryphon, E. and Soares jr, J.H. 2001. The effect of phytrase on apparent digestibility of four practical plant feedstuffs fed to striped bass, Morona saxatilis. Aquacult. Nutr. 7, 161-167

Papoutsoglou, S.E. and Tziha, G. 1996. Blue tilapia (*Oreochromis aureus*) growth rate in relation to dissolved oxygen concentration under re-circulated water conditions. Aquacult. Eng. 15(3), 181-192

Reigh, R.C., Braden, S.L. and Craig, R.G. 1990. Apparent digestibility coefficients for common feed stuffs in formulated diets for red swamp cray fish, *Procambarus clarkii*, Aquaculture 84, 321-334

Robaina, L., Izquierdo, M.S., Moyano, F.J., Socorro, J., Vergara, J.M., Montero, D. and Fernandez-Palacios, H. 1995. Soybean and lupin seed meals as protein sources in diets for gilthead seabream (*Sparus aurata*): nutritional and histological implications. Aquaculture 130, 219-233

Sales, J. and Britz, P.J. 2001. Evaluation of different markers to determine apparent nutrient digestibility coefficient of feed ingredients for South African abalone (*Haliotis midae* L.). Aquaculture 202, 113-129

Sugiura, S.H., Dong, F.M., Rathbone, C.K. and Hardy, R.W. 1998. Apparent protein digestibility and mineral availability in various feed ingredients for salmonid feeds. Aquaculture 159, 177-202

Sullivan, J.A. and Reigh, R.C. 1995. Apparent digestibility of selected feed stuffs in diets for hybrid stripped bass (*Morone saxatilis* × *Morone chrysops*). Aquaculture 138, 313-322

Tacon, A.G.J. and Rodrigues, A.M.P. 1984. Comparison of chromic oxide, crude fiber, polyethyl and Acid-Insoluble Ash as dietary markers for the estimation of apparent digestibility coefficients in *Rainbow trout*. Aquaculture 43, 391-399

Tantikitti, C. and Chimsung, N. 2001. Dietary lysine requirement of freshwater catfish (*Mystus nemurus* Cuv. & Val.). Aquacult. Res. 32, 135-141

Van Den Berg, G.W. and De La Noue 2001. Apparent Digestibility comparison in rainbow trout (Oncorhyncus mykiss) assessed using three methods of faeces collection and three digestibility markers. Aquacult. Nutr. 7, 237-245

Van Dyke, J.M. and Sutton, D.L. 1977. Digestion of duckweed (Lemna sp.) by the grass carp (*Ctenopharyngodon idella*). J. Fish Biol. 11, 273-278

Weatherup, R.N. and McCracken, K.J. 1998. Comparison of estimates of digestibility of two diets for rainbow trout, *Oncorhyncus mykiss* (Walbaum) using two markers and two methods of faeces collection. Aquacult. Res. 29, 527-533

Weerasinghe, V., Hardy, R.W., and Haard, N.F. 2001. An in vitro method to determent phosphorus digestibility of rainbow trout *Oncorhynchus mykiss* (Walbaum) feed ingredients. Aquacult. Nutr. 7, 1-9

Yamamoto, T., Shima, T., Furuita, H., Suzuki, N., Sanchez-Vazquez, F.J. and Tabata, M. 2001. Self-selection and feed consumption of diets with a complete amino acid composition and a composition deficient in either methionine or lysine by rainbow trout, *Oncorhynchus mykiss* (Walbaum). Aquacult. Res. 32, 83-91

Zhu, S., Chen, S., Hardy, R.W. and Barrows, F.T. 2001. Digestibility, growth and excretion response of rainbow trout (*Oncorhynchus mykiss* Walnaum) to feeds of different ingredient particle sizes. Aquacult. Res. 32, 885-893

Chapter Nine

Summary

Summary

Introduction

About 80 countries inhabited by 40% of the world's population were suffering from serious water shortages by 1990 and it is estimated that within 25 years two-thirds of the world population will suffer from water stress (UNEP, 2002). Extrapolations under the "business as usual" scenario show that by 2020, water consumption is expected to increase by 40%, and 17% more water will be required for food production to meet the need of the growing human population. Agriculture accounts for consumption of more than 70% of all water abstractions from rivers, lakes and ground water. Irrigated agriculture provides 40% of the world human food production. Despite the increased percentage of people served with safe drinking water supply from 79% in 1990 to 82% in 2000, 1.1 billion people still lack access to safe drinking water and 2.4 billion people have no access to sanitation services; most of them are in Africa and Asia (UNEP, 2002). Sanitation coverage in rural areas is less than in urban settings. 80% of people lacking sanitation live in rural areas (WHO and UNICEF, 2000). As a consequence, a clear need for extending sustainable and safe sanitation services to small towns and rural communities, has emerged in the last decade. The objective of national and international policies was to identify the options and requirements for initiating and accelerating expansion of such systems in small communities especially in water scarce countries. The Food and Agriculture Organisation (FAO) of the United Nation (UN) has drawn great attention to secure food for people with affordable prices now and in the future. The expected increase in water demand due to increase in human population and per capita domestic water consumption, will negatively affect irrigated agriculture. Also deterioration and lose of fertility of soil due to intensive use of artificial fertilisers and pesticides will negatively affect the productivity of irrigated lands as well. Discharge of municipal sewage and drainage water in water streams enhances eutrophication of water bodies with subsequent deterioration in water quality. Deterioration of the quality of surface water reduces natural stock of fish in rivers and lakes with severe impacts on people's income and food security in the nearby societies. Intensive use of chemical fertilisers and pesticides increase contamination of the ground water aquifers. Severe contamination of surface and ground water negatively affects efficiency of water treatment plants, which derive influent water from these contaminated sources. Better management of natural resources via implementing affordable wastewater treatment technology with nutrients recovery and water recycling could safeguard, water resources and soil fertility with subsequent increased food security for the society.

Sustainable development has been well-defined by the FAO (1991) as "the management and conservation of the natural resources and the orientation of technological and institutional change in such a manner as to ensure the attainment and continued satisfaction of human needs for present and future generations". Such sustainable development has to be environmentally non-degrading, technically appropriate, economically feasible and socially acceptable. Conceptually, human wastes and drainage are resources in a wrong place since it flows to the environment mostly without prior treatment. Nutrients in these wastewaters have an economic value and should be treated in such way to achieve management of natural resources and implementation of sustainable development strategies. To achieve this approach wastewater should be considered as a resource and its management represents an integral part in the whole process of water resources and nutrients flow management strategies.

Decentralised sanitation and reuse concepts could be considered as environmentally sustainable options for developing countries in particularly the arid and semiarid regions. Anaerobic treatment of domestic wastewater followed by duckweed ponds provides a technology that is simple, low cost, and affordable to be locally financed and managed under proper operating conditions. The duckweed pond system is based on solar energy and there is no need for external power supply. Macrophyte ponds stocked with duckweed and the role of duckweed in the treatment process and nutrient recovery, have been studied (Al-Nozaily *et al.*, 2000a; Al-Nozaily *et al.*, 2000b, Zimmo *et al.*, 2000; Smith and Moelyowati, 2001; Cheng *et al.*, 2002). In these ponds, duckweed and bacteria are the dominant organisms, which catalyse the process of purification. Duckweed contains considerable amounts of protein and represents a valuable marketable by-product, which could generate cash income. Duckweed is the smallest flowering aquatic plant and has no stem or leaf structure. This unique feature makes that the biomass is low in fibre content and it provides a good quality feed for tilapia. Therefore duckweed biomass provides a much more attractive fish feed compared to other aquatic weeds such as water hyacinth, which is low in protein and high in fibre content (El-Sayed, 2003). Duckweed might also be better than oilcakes, which contain some anti-nutritional factors (Singh *et al.*, 2003). The application of duckweed (*Lemna spp.*) in carp polyculture ponds had positive effects on the growth performance (Azim and Wahab, 2003).

Objectives

In this thesis an Up-flow Anaerobic Sludge Blanket (UASB) reactor in combination with duckweed and tilapia ponds were investigated as an integrated system for sewage treatment and reuse in fish aquaculture. The UASB-duckweed-tilapia system is a natural system, which provides a sustainable approach to fish aquaculture since domestic sewage is the only input source for the system. Up to now, the potential of such an integrated system to the advancement of sustainable freshwater aquaculture has not been confirmed, since very little research has been done in this area. The general objective of this research is the evaluation of the UASB-duckweed ponds for domestic wastewater treatment and fish feed production. A number of specific research questions are addressed. These specific objectives are:

-To monitor the efficiency of the UASB-duckweed pond system for sewage treatment and nutrient recovery and to evaluate the effect of seasonal temperature changes on the treatment efficiency and duckweed production.

-To study the effect of feeding rate and feed form on water quality and fish yield in a duckweed-based tilapia aquaculture system.

-To evaluate the use of fresh duckweed in tilapia pond culture by comparison with wheat bran or fishmeal-based commercial feed.

-To study the effect of water quality (freshwater, treated sewage from the UASB-duckweed ponds and settled sewage) on fish yield in tilapia ponds.

-To study the effects of microbial quality and other water quality parameters on microbial contamination of tilapia reared in different faecal contaminated ponds.

-To study the effect of chronic ammonia toxicity on the growth performance of duckweed-fed tilapia.

-To assess the digestibility coefficients of duckweed (fresh and dry) to be used as fish feed ingredients in intensive culture of tilapia.

Summary of results and discussion

Chapter 2 focuses on the UASB-duckweed pond system for the treatment of domestic sewage. The effect of temperature on the system efficiency during summer and winter and the fate of nitrogen in duckweed ponds were studied. The experiment was conducted in a pilot-scale system consisting of a UASB (40 l   capacity) and three duckweed ponds in series (480 l each). The reactor was fed with domestic sewage at a hydraulic retention time of 6 hours to provide 160 l treated effluent, daily. 96 litres of UASB effluent was pumped daily to the three duckweed ponds, providing 5 days HRT in each. The ponds were stocked with *Lemna gibba* at 600 g fresh biomass per square meter. The treatment system was intensively monitored for one year. The nitrogen mass balance was determined in the three duckweed ponds by calculating nitrogen recovery, ammonia volatilisation and nitrogen sedimentation. Nitrogen recovery by duckweed was calculated by monitoring daily growth rate of duckweed and composition of plant biomass. Ammonia volatilisation was measured by trapping the emitted gas in 2% boric acid solution with subsequent analysis for ammonia. Analysis of accumulated sediment for dry matter, nitrogen and phosphorus content was done. In spite of the significant reduction in the UASB removal efficiency of COD, BOD and TSS in winter (79%, 82% and 83% for COD, BOD and TSS, respectively in summer compared to 70%, 73% and 73% in winter) the efficiency of the complete system was not significantly affected (93%, 95%, and 91% in summer in comparison to 92%, 93% and 91% in winter). Final effluent contained 49 mg COD/l, 14 mg BOD/l and 32 mg TSS/l in summer in comparison to 73 COD/, 25 mg BOD/l and 31 mg TSS/l in winter. COD, BOD and TSS removal efficiency in the duckweed ponds was higher in winter in comparison to the summer period due to the presence of large numbers of daphnia and copepods in the pond effluent in summer. Nutrient removal was significantly reduced in winter. In summer effluent values of ammonia, TKN and total phosphorus were 0.4 mgN/l, 4.4 mgN/l and 1.1 mgP/l, with removal efficiencies of 98%, 85% and 78%, respectively. In winter, effluent values for the same parameters were 10.4 mgN/l, 12.2 mgN/l and 2.7 mgP/l, with removal efficiencies of 39%, 53% and 57%, respectively. This shows that the nutrient removal processes are more temperature dependent than the BOD and COD removal mechanisms. In summer, the faecal coliform count was reduced from 2.9×10^8 cfu/100ml in the sewage to 8.9×10^7 cfu/100ml in the UASB effluent and 4.0×10^3 cfu/100ml in the final pond effluent. In winter the values were 1.1×10^9, 3.2×10^8 and 4.7×10^5 cfu/100ml in the raw sewage, UASB effluent and final pond effluent, respectively. Microbial quality of UASB effluent and raw sewage indicated that 63% and 73% reduction in the faecal coliform count was achieved in summer and winter, respectively. This is in agreement with 70% reduction reported by Lettinga *et al.* (1992). The results of this study show the effective removal of faecal coliform in duckweed ponds during the summer in comparison to the winter season. The removal mechanisms might be adsorption on duckweed biomass, sedimentation, decay and grazing by protozoan.

The summer nitrogen mass balance showed that 80.5% of total nitrogen removal in ponds was achieved by plant uptake, while 4.7% was removed by sedimentation and 14.8% by de-nitrification. Ammonia volatilisation was negligible in ponds. Nitrogen uptake by duckweed ranged between 4 and 4.9 KgN/ha/d in summer and 1.2 to 1.5 KgN/ha/d in winter. The total P recovered by duckweed ranged from 0.86 to 0.97 Kg P/ha/d in summer and from 0.27 to 0.32 Kg P/ha/d in winter. These results show the negative effects of temperature on the growth rate of duckweed. The duckweed production rate was 126-139 kg dry matter/ha/d in summer and 31-36 kg dry matter/ha/d in winter. In warm seasons, duckweed production rates amounted to 33 ton dry matter per hectare per 8 months (warm seasons). The market price (2002) of this could be 16,830 Egyptian pound (4,360 $) assuming the price of duckweed is equal to the local price of the cheapest local feed ingredient (wheat bran). The dry matter content of duckweed biomass was 4.3-5.4% with 20-25.7% protein content and 0.68-0.9% P. This chapter concluded that the UASB-duckweed pond system is technically appropriate for sewage treatment in peri-urban areas, small communities and rural areas and provides marketable by-products (duckweed biomass).

The nutritional value of duckweed biomass harvested from the duckweed ponds in raising tilapia was evaluated using batch experiment (**Chapter 3**). Effects of daily feeding rate and mode of application of duckweed (fresh and dry) on growth performance of Nile tilapia (*Oreochromis niloticus*) and water quality were investigated. The duckweed (*Lemna gibba*) was evaluated as sole feed source for Nile tilapia at a temperature range of 16-25 °C. Two batches of juveniles with initial mean body weight of about 8 g and 20 g were tested using fishponds with 30 litre working volume and 0.25 m^2 surface area. The ponds were stocked with five fish per tank and 15 litre of water was renewed weekly. Daily feeding rates in the five treatments were 25% (dried), 50% (dried), 10% (fresh), 25% (fresh) and 50% (fresh), respectively. The feeding rates were calculated based on the fresh weight of both duckweed and fish. The duckweed used in treatments 1 and 2 was dried in an oven at 105 °C. The results of growth performance showed no significant (p >0.05) difference between the specific growth rates (SGRs) of the two batches. The SGRs were 0.24, 0.24, 0.25, 0.55 and 0.51 in the treatments 1, 2, 3, 4 and 5, respectively. Statistical analysis showed significantly (p < 0.01) higher SGR for tilapia fed with fresh duckweed (treatment 4 and 5) than for those of the parallel treatments fed with dry duckweed, (treatment 1 and 2). This could be attributed to the moisture content of fresh duckweed, which might have a role in enhancing the digestibility of the feed. Also Tacon *et al.* (1995) reported that feed pellet with 39% moisture content resulted in better SGR and FCR and less solid wastes compared to feed pellets with 12% moisture content in culture of sea bass (*Lates calcarifer*) and grouper (*Epinephelus tauvina*). Guerrero (1980) reported similar results in cage culture of tilapia where significant higher SGR and net yield were obtained in case of moist feed pellets in comparison to dry feed pellets. Also sewage bacteria attached to fresh duckweed might have enhanced microbial decomposition of duckweed in the fish intestine since microbial decomposition of duckweed (*Lemna gibba*) was three folds higher in the presence of sewage bacteria. The time required to lose half of the organic matter of duckweed in the presence of sewage bacteria was one third of that in the absence of sewage bacteria (Szabó *et al.*, 2000). In addition, the drying process might cause loss of some essential elements and this may also explain the lower SGR observed in the treatments with dry duckweed. The large volume of fresh duckweed had no negative effect on feed intake, which might be attributed to the low feeding rates. In case of fresh duckweed, the SGR of tilapia fed at 10% feeding rate had a significant ($p < 0.01$) lower value than in 25% and 50% feeding rate, while there was no significant ($p > 0.05$) difference between 25% and 50%. The low optimal feeding rate (25%) observed in this experiment might be due to, the low temperature range (16-25 °C), which negatively affects metabolic activity of fish since the optimal temperature range for *Oreochromis niloticus* is 28-30 °C (Balarin and Haller, 1982). In case of dry duckweed no significant ($p > 0.05$) differences were detected between the 25% and 50% feeding rate. The feed conversion ratios (FCRs) for treatment 1 to 5 were 4.6, 9.2, 1.7, 1.9 and 4.1 in the 20g/fish batch of tilapia and 4.4, 8.7, 1.6, 1.9 and 4.2 in the 8g/fish batch of tilapia, respectively. The protein efficiency ratios (PERs) were 0.9, 0.45, 2.5, 2.1 and 1.0 in the treatments 1, 2, 3, 4 and 5, respectively. Results of the nitrogen mass balance showed that the nitrogen content of dry sediment (solid waste) ranged between 4% and 10.5%. In case of optimal feeding rate (25% fresh duckweed) the value was 4.5%. This amount of nitrogen represented 8% of the dietary nitrogen input while nitrogen recovery by fish represented 26-28% of the total nitrogen input. The major part of N loss was due to effluent discharge (10% and 25% feeding rate) or due to storage in the sediment (50% feeding rate).

Use of both treated sewage and duckweed biomass from the UASB-duckweed pond system was evaluated in rearing Nile tilapia (*Oreochromis niloticus*) in ponds (**Chapter 4**). Comparison of fresh duckweed grown on anaerobically treated sewage and a local fish feed ingredient, wheat bran, as two different supplementary feeds was investigated. Three different water sources, including freshwater, treated effluent of the UASB-duckweed pond system and settled sewage were investigated. The experiment was conducted using five ponds with 1 m^2 surface area. Each pond was stocked with 10 juveniles of Nile tilapia weighing about 20 g. The nutritional value of duckweed was compared with wheat bran by applying them separately as the only source of feed using treated-sewage or freshwater. Duckweed feeding rate was set at 25g fresh duckweed per 100 g fish live weight. The wheat bran fed ponds received the same amount of dry matter as the duckweed ponds. Parallel to these incubations, another pond received settled raw sewage as a sole source of feed. The results of growth performance showed that, in case of freshwater ponds, the SGR of tilapia fed on fresh duckweed was significantly ($p < 0.01$) higher than the SGR of those fed on wheat bran. This might be due to the significantly higher protein content of duckweed in comparison to wheat bran. No significant difference ($p > 0.05$) was observed between the treatments in case of treated effluent-fed ponds. This may be because spawning in the duckweed ponds occurred earlier than in the wheat bran ponds (75 days later). This may have negatively affected SGR and this resulted in non-significant differences with the wheat bran fed pond. In the wheat bran fed ponds the SGR of tilapia reared in the treated sewage pond (TWP) was significantly ($p < 0.01$) higher than the SGR of those reared in freshwater pond (FWP). In duckweed fed ponds, the SGR of tilapia reared in treated sewage pond

(TDP) was significantly lower (p <0.05) than the SGR of those in the freshwater pond (FDP). This might be attributed to the early spawning in TDP, as mentioned earlier. Net fish yield were 11.8, 9.6, 8.9 and 6.4 ton/ha/y in TDP, FDP, TWP and FWP. The lowest net yield was in settled sewage-fed fishpond (SSP) and was -0.16 ton/ha/y. The poor result obtained in SSP was attributed to the high mortality in this pond, reaching up to 60% in the adult fish and 38% in the fry. This mortality might be due to chronic toxicity of un-ionised ammonia, which was 0.12 mg N/l on average during the experimental period. Fish mortality was observed within the last two months of the experiment (autumn), during which the average un-ionised ammonia concentration was 0.45 mg N/l. The treated sewage fed ponds provided higher net yield in comparison to freshwater ponds. This could be attributed to the role of nutrients present in treated sewage in primary productivity enhancement, which contributes to satisfy some of the nutrient requirements of fish (Tidwell *et al.*, 1995; Agustin, 1999). Presence of daphnia and copepods in the treated sewage might also have contributed to the high yields in treated sewage fed ponds (McEvoy *et al.*, 1998; Warburton *et al.*, 1998; Kibria *et al.*, 1999; Lienesch and Gophen, 2001).

In **chapter 5** effects of prolonged exposure to sub-lethal un-ionised ammonia concentrations on the growth performance of juveniles of tilapia (*Oreochromis niloticus*) fed on fresh duckweed (*Lemna gibba*) was investigated. The experiment was conducted in duplicates over 75 days using juveniles with mean body weight 20g. Five nominal, total ammonia nitrogen concentrations (control, 2.5, 5, 7.5 and 10 mg N l^{-1}) were established as treatment groups. Statistical analysis of the specific growth rate (SGR) showed no significant (p > 0.05) differences between the SGR (0.71) of tilapia in the control (0.004 mg UIA-Nl-1) and the SGR (0.67) of those exposed to 0.068 mg UIA-N l^{-1}. The SGR of tilapia exposed to un-ionised ammonia nitrogen over 0.068 mg UIA-N l^{-1} (0.144, 0.262 and 0.434 mg UIA-N l^{-1}) was significantly reduced (p< 0.01). The maximum no observable effect concentration was 0.068 mg UIA-N l^{-1}, while the lowest observable effect concentration was 0.144 mg UIA-N l^{-1}. Increasing the un-ionised ammonia concentration increased the feed conversion ratio (FCR). At 0.144 mg UIA-N l^{-1} the FCR increased to 1.6 times the value observed in the control, while at 0.262 mg UIA-N l^{-1} the FCR increased 2.7 fold. The protein efficiency ratio (PER) was also negatively correlated with un-ionised ammonia concentration above 0.068 mg UIA-N l^{-1}. At 0.434 mg UIA-N l^{-1} the FCR had increased 4.3 fold. This study concluded that for raising *Oreochromis niloticus* in fishponds, the UIA-N has to be maintained below 0.1 mg UIA-N l^{-1}. In this experiment no fish mortality was detected in any of the treatment groups even at the highest concentration (0.43 UIA-N l^{-1}). This result seems to be contrary to the data obtained in chapter 4 where in settled sewage pond fish mortality was observed at UIA-N of 0.45 mg l^{-1}. This might be attributed to the cumulative negative effects of other water quality parameters like sewage bacteria count and low dissolved oxygen at dawn in SSP. These parameters might have synergistic effects to the toxicity of UIA-N in SSP.

In **chapter 6** microbial quality of tilapia reared in four faecal contaminated fishponds, as described in chapter 4, was investigated. In TDP the ponds received two sources of contamination, treated sewage and fresh duckweed grown on anaerobically treated sewage. In TWP and FDP there was only one source of contamination in the form of treated sewage in the first and fresh duckweed in the latter. The last pond (SSP) received only settled sewage as source of water and indirect feed. The treated domestic sewage had an average count of faecal coliform of 4×10^3 cfu/100ml, while the average number of faecal coliform in settled sewage was about 2.1×10^8/100ml. The number of faecal coliform on the duckweed biomass ranged from 4.1×10^2 to 1.6×10^4 cfu/g fresh weight. Monitoring the water microbial quality showed an average count of faecal coliform of 2.2×10^3, 1.7×10^3, 1.7×10^2 and 9.4×10^3 cfu/100ml in TDP, TWP, FDP and SSP, respectively. Statistical analysis showed no significant ($p > 0.05$) differences in faecal coliform counts between both treated sewage fed ponds (TDP and TWP) and SSP. The FDP had significantly lower faecal coliform count than the treated sewage-fed ponds ($p < 0.01$) and SSP ($p < 0.05$). Investigation of microbial quality of tilapia raised in these ponds indicated that all tissue and organ samples were contaminated with faecal coliform except the muscle tissues, which had no faecal contamination at all. Ranking of faecal coliform contamination showed a decrease in the order intestine > gills > skin > liver. In case of TDP no significant difference ($p > 0.05$) in microbial quality, was observed neither, between the intestine and gills nor between the skin and liver. In case of TWP significantly higher ($p < 0.05$) faecal coliform contamination was observed in the intestine compared to gills, but no difference was noticed between the skin and the liver ($p > 0.05$). In FDP significantly higher contamination occurred in the intestine compared to the other organs ($p < 0.01$). In SSP there was no significant difference between the intestine and gills ($p > 0.05$) but levels in these organs were higher than those in the liver and the gills ($p < 0.05$). In spite of absence of any significant differences between the pond water microbial quality in SSP and treated sewage ponds, significantly higher contamination was observed in fish organs in the latter. This might be attributed to other water quality parameter such as UIA-N and nitrite nitrogen that were significantly higher in SSP. This poor water quality in SSP resulted in faecal coliform levels in fish organs of about one $\log_{10}$ higher than in treatments with good water quality. Pre-treatment of sewage is therefore recommended before use in aquaculture not only for faecal colifrom reduction but also ammonia and nitrite.

In this chapter (**Chapter 7**) fresh duckweed was compared with local fishmeal-based diet, while using treated sewage and de-chlorinated city tap water as a water source. Fresh duckweed (*Lemna gibba*) harvested daily from duckweed sewage stabilisation ponds was applied in culture of Nile tilapia (*Oreochromis niloticus*) with 23 g average initial body weight. The daily feeding rate in duckweed-fed fishponds was 25 g fresh duckweed per 100 g of live fish weight. The fishponds fed with commercial feed received equal amounts of protein as in the duckweed-fed ponds. The experiment was conducted in four ponds, each of 1 m^2 and 480 l, within a temperature range of 26-34 °C for a 120 days growth period. In the freshwater ponds the juveniles of tilapia fed on the commercial feed had significantly ($p < 0.01$) higher SGR than in duckweed-fed ponds. This might be attributed to deficiency in some essential elements and additives in duckweed like cysteine and methionine (Hammouda *et al.*, 1995). These additives and essential elements might be supplied by natural productivity in case of treated sewage fed ponds since in these ponds no significant difference between the duckweed and commercial feed ponds was observed. D'Abramo and Conklin (1995) reported that the natural biota could significantly contribute to satisfy some of the nutrients such as vitamins, cholesterol, phospholipids and minerals, and this was supported by results of Omondi *et al.* (2001). The biomass of daphnia and copepods that was observed in treated effluent might also satisfy deficiencies of duckweed, since these crustaceans have excellent feed value for pisciculture (Bogatova *et al.*, 1973; Proulx and de La

Noue, 1985). The positive effects of treated effluent have also been detected in comparison to the use of freshwater. The SGR and fish yield were higher in treated sewage fed ponds, both in duckweed ponds and commercial feed ponds. Fish net yields were 15.7, 16.8, 8.5 and 11.2 ton/ha/y, in TDP, TCP, FDP and FCP, respectively.

In **Chapter 8** apparent digestibility coefficients of duckweed (*Lemna minor*) were measured for Nile tilapia (*Oreochromis niloticus*). Tilapia was fed on four iso-nitrogenous treatment diets in addition to the control diet. The control diet contained fishmeal (35%), corn grain (29%), wheat grain (20%), wheat bran (10%), fish oil (3%), diamol (2%) and premix (1%). The effect of partial replacement of control diet components with 20% and 40% dry duckweed based on dry matter in diets of treatment 1 (T1) and treatment 2 (T2) was investigated. In treatment 3 (T3) and treatment 4 (T4), fresh duckweed was added to replace 20% and 40% of the control diet components based on dry matter basis, respectively. During the 49 days experimental period faeces were collected and the water quality was monitored. At the end of the trial growth performance, digestibility coefficients, carcass body composition, faecal recovery percentage and nitrogen mass balances were assessed. Specific growth rates of tilapia were 1.51 ± 0.07, 1.38 ± 0.03, 1.31 ± 0.06, 1.44 ± 0.02 and 1.33 ± 0.05, in control and treatments 1 to 4. There was no significant difference between SGR for the control and T3 while the other treatment groups had significantly lower SGR. All of the treatment diets provided good values for FCR (0.98-1.1) and PER (2.49-2.78). The average values of FCR and PER of the control diet were 0.91 and 3, respectively. Except for T2, which had a higher significant faeces recovery percentage ($p<0.05$), there were no significant differences between the control and the treatments. Faeces recovery in T1 was significantly lower ($p<0.05$) than in the other treatments. There was no significant difference between the dry matter content of faeces except in treatment 4, which had a significantly lower value in comparison to the control and T2. The dry matter digestibility (DMD) ranged from 61.8% in treatment 4 to 85.2% in control and the range for the treatment diets was 64%-71%. All the treatment diets had high protein digestibility (78%-92%) and high-energy digestibility (78.1%-90.7%). It is therefore possible to use duckweed as replacement of fishmeal and other plant ingredients in fishmeal-based diets at 20% and 40% inclusion. Digestibility coefficients of dry and fresh duckweed showed DMD of 66.8, 63.3, 45.8 and 28.3 in the treatments 1 to 4, respectively. These results, even in case of 40% fresh duckweed, are comparable to the reported range (27.5%-50%) for wheat middling, wheat flour, corn grain and yellow maize (Sullivan and Reigh, 1995; Fagbenro, 1996). Protein digestibility coefficients (PDC) values of duckweed, were 78.4, 79.9, 77.6 and 75.9 while energy digestibility (ED) values were 59.8, 60.9, 64.5 and 58.4 in T1, T2, T3 and T4, respectively. For 20% duckweed inclusion there was no significant difference between dry and fresh duckweed for the apparent digestibility coefficients of dry matter, protein, fat, phosphorus and energy. However, in case of 40% inclusion the fresh duckweed had significantly lower digestibility coefficient values for dry matter, protein, fat and phosphorus. These results might be attributed to dilution of the enzyme concentration by duckweed moisture in case of 40%. The milling process might also have positive effects on the apparent digestibility of dry duckweed since destruction of cell walls of duckweed exposes the internal tissues to enzyme activity. Decreasing the particle size of duckweed also increases specific surface area for enzyme reactions with as a result improvement of the apparent digestibility coefficients. Body composition showed that tilapia fed on diets with duckweed have significantly ($p<0.05$) higher ratio of phosphorus and protein and lower (p<0.05) lipid content. This might be attractive for the consumers.

In chapter 3, growth performance of tilapia fed on fresh duckweed was better than those fed on dry duckweed. The results of apparent digestibility coefficients of dry and fresh duckweed in this chapter seem to be contrary to the results in chapter 3. This may be explained by the following reasons. First, the positive effects of the milling process (Zhu et al., 2001) in the digestibility trial may have compensated the negative effects of drying to some extent comparing to the non-processed dry duckweed in chapter 3. Secondly in chapter 3 the duckweed species was *Lemna gibba*, which is characterised by larger fronds, in comparison to *Lemna minor* used in the digestibility trial. The large frond makes it difficult for tilapia to swallow the plant without chewing while small size of *Lemna minor* (digestibility trial) makes it easy for the fish to swallow the plant without chewing. The chewing process increases the digestibility via destruction of cellulose in the cell wall of the plant. Thirdly the sewage bacteria attached to *Lemna gibba* (chapter 3) may have had a positive effect on the digestion process.

Conclusions

Efficiency of a UASB-duckweed pond system in removing COD, BOD and TSS was sufficient during the whole year, while nutrient recovery in duckweed ponds was significantly reduced at lower temperatures (13-20 °C) since the nutrient recovery is a duckweed growth mediated process. Faecal coliform removal in duckweed ponds was negatively affected by the decline in temperature therefore probably removal mechanisms were affected by temperature, nutrients availability and duckweed harvesting rate. In general the UASB-Duckweed pond system could be technically appropriate, socially acceptable and environmentally sustainable for domestic wastewater treatment in rural areas and small communities.

The use of the UASB-duckweed pond system in integrated sewage-tilapia aquaculture in ponds is superior to and safer than the direct use of sewage. Fresh duckweed (*Lemna gibba*) is superior (higher fish yield) to wheat bran as supplementary feed and has equal nutritional value as fishmeal-based diets, especially in treated sewage-fed ponds. The use of treated domestic sewage from UASB-duckweed pond, as replacement of freshwater is sustainable since fertilisation of tilapia ponds with treated sewage from UASB-duckweed ponds increases the fish yield significantly without further resource input.

Water quality parameters such as ammonia and nitrite in fishponds may affect the level of bacterial contamination of fish organs, probably due to the effect of water quality on the fish immune response. Therefore it is recommended to pre-treat sewage before using it in fish aquaculture. The low DO concentration and microbial quality of water might have synergistic effect on the chronic ammonia toxicity. To avoid any negative impacts of un-ionised ammonia nitrogen (UIA-N) on the growth performance of tilapia the UIA-N concentration in the pond should be maintained lower than 0.1 mg UIA-N l^{-1}.

Digestibility coefficients of duckweed (*Lemna minor*) are better than some conventional plant ingredients used in fish feed. The results of digestibility coefficients of diets containing 20% and 40% inclusion of dry duckweed and diets with 20% inclusion of fresh duckweed encourage the use of duckweed in fish feed manufacturing. The use of duckweed in dry or fresh form could replace more costly and often scarcely available fishmeal-based diets.

Recommendations

Since the nutrient recovery in the UASB-duckweed pond system is significantly reduced by lower temperature (13-20 °C) effluent storage in a reservoir is recommended during the winter to be reused in summer and in case of direct reuse of the treated effluent during the winter in agriculture or aquaculture ammonia nitrogen and TKN have to be taken into account.

References

Agustin, C.P. 1999. Role of natural productivity and artificial feed in the growth of freshwater prawn, *Macrobrachium borelli* (Nobili, 1896) cultured in enclosures. J. Aquacult. Trop. 14(1), 47-56

Al-Nozaily, F., Alaerts, G. and Veenstra, S. 2000. Performance of duckweed-covered sewage lagoon. 1. Oxygen balance and COD removal. Water Res. 34(10), 2727-2733

Al-Nozaily, F., Alaerts, G. and Veenstra, S. 2000a. Performance of duckweed-covered sewage lagoon. 2. Nitrogen and phosphorus balance and plant productivity. Water Res. 34(10), 2734-2741

Azim, M.E. and Wahab, M.A. 2003. Development of a duckweed-fed carp polyculture system in Bangladesh. Aquaculture 218, 425-438

Balarin, J.D. and Haller, R.D. 1982. The intensive culture of tilapia in tanks, raceways and cages. In: James F. Muir & Ronald J. Roberts (Eds.), Recent advances in aquaculture. Croom Helm Ltd, London. pp. 265-355.

Bogatova, I.B., Schhesbina, M.A., Ovinnikova, V.V. and Tagisova, N.A. 1973. The chemical composition of certain planktonic animals under different growing conditions. Hydrobiol. J. 7, 39-43

Cheng, J., Bergamann, B. A., Classen, J.J., Stomp, A.M. and Howard, J.W. 2002. Nutrient recovery from swine lagoon water by *Spirodela punctata*. Biores. Technol. 81, 81-85.

D'Abramo, L.R. and Conklin, D.E. 1995. New development in the understanding of the nutrition of penaeid and caridean species of shrimp. Proceedings special session on shrimp farming, aquaculture 95, World Aquaculture Society, Louisiana, USA, pp. 95-107

El-Sayed, A.M. 2003. Effects of fermentation methods on the nutritive value of water hyacinth for Nile tilapia *Oreochromis niloticus* (L.) fingerlings. Aquaculture 218, 471-478

Fagbenro, O.A. 1996. Apparent digestibility of crude protein and gross energy in some plant and animal-based feedstuffs by *Clarias isheriensis* (Siluriformes: Clariidae) (Sydenham 1980). J. Appl. Ichthyol. 12, 67-68

Guerrero, R.D.III, 1980. Studies on the feeding of *Tilapia nilotica* in floating cages. Aquaculture, 20, 169-175

Hammouda, O., Gaber, A. and Abdel-Hameed, M.S. 1995. Assessment of the effectiveness of treatment of wastewater-contaminated aquatic systems with *Lemna gibba*. Enz. and Micro. Technol. 17: 317-323

Kibria, G., Nugegoda, D., Fairclough, R., Lam, P. and Bradley, A. 1999. Utilization of wastewater-grown zooplankton: Nutritional quality of zooplankton and performance of silver perch bidyanus bidyanus (Mitchell 1838) (Teraponidae) fed on wastewater-grown zooplankton. Aquacult. Nutr. 5, 221-227

Lettinga, D., de Man, A., van der Last, A. R. M., Wiegan, W., van Knippenberg, K., Frings, J. and van Bauren, J. C. L., 1992. Anaerobic treatment of domestic sewage and wastewater.

"Proceeding of first Middle East conference on water supply and sanitation for rural areas". Pp. 164-170, Cairo, Egypt.

Lienessch, P.W. and Gophen, M. 2001. Predation by inland silversides on exotic cladoceran, daphnia lumholtzi, in lake Texoma, USA. J. Fish Biol. 59, 1249-1257

McEvoy, L.A., Naess, T., Bell, J.G. and Lie, Ø. 1998. Lipid and fatty acid composition of normal malpigmented Atlantic halibut (*Hippoglossus hippoglossus*) fed enriched Artemia: A comparison with fry fed wild copepods. Aquaculture 163, 237-250

Omondi, J.G., Gichuri, W.M. and Veverica, K. 2001. A partial economic analysis for Nile tilapia (*Oreochromis niloticus* L.) and sharptoothed catfish *Clarias gariepinus* (Burchell 1822) polyculture in central Kenya. Aquacult. Res. 32, 693-700

Proulx, D. and De La Noue, J. 1985. Harvesting *Daphnia magna* grown on urban tertiarily treated effluents. Water Res. 19(10), 1319-1324.

Singh, K., Garg, S.K., Kalla, A. and Bhatnagar, A. 2003. Oilcakes as protein sources in supplementary diets for the growth of *Cirrhinus mrigala* (Ham.) fingerlings: laboratory and field studies. Biores. Technol. 86, 283-291

Smith, M.D. and Moelyowati, I. 2001. Duckweed based wastewater treatment (DWWT): design guidelines for hot climates. Water Sci. Technol. 43(11), 291-299

Sullivan, J.A. and Reigh, R.C. 1995. Apparent digestibility of selected feedstuffs in diets for hybrid stripped bass (*Morone saxatilis* × *Morone chrysops*). Aquaculture, 138, 313-322

Szabó, S., Braun, M., Nagy, P., Balazsy, S. and Reisinger, O. 2000. Decomposition of duckweed (*Lemna gibba*) under axenic and microbial conditions: flux of nutrients between litter water and sediment, the impact of leaching and microbial degradation. Hydrobiologia 434, 201-210.

Tacon, A.G.J., Phillips, M.J. and Barg, U.C. 1995. Aquaculture feeds and the environment: the Asian experience. Water Sci. Technol. 31(10), 41-59.

Tidwell, J.H., Webster, C.D., Sedlacek, J.D., Weston, P.A., Knight, W.L., Hill, Jr S.J., D'Abramo, L.R., Daniels, W.H., Fuller, M.J. and Montanez, J.L. 1995. Effects of complete and supplemental diets and organic pond fertilisation on production of *Macrobracheium rosenbergii* and associated benthic macroinvertebrate populations. Aquaculture, 138, 169-180

UNEP 2002. State of the environment and policy retrospective: 1972-2002. In Global Environment Outlook 3 Past, Present and Future Perspectives (Eds. Production team of UNEP), UNEP, London, UK.

Warburton, K., Retif, S. and Hume, D. 1998. Generalists as sequential specialists: diets prey switching in juvenile silver perch. Environ. Biol. Fishes 51, 445-454

WHO and UNICEF 2000. Global Water Supply and Sanitation Assessment 2000 Report, pp 80

Zhu, S., Chen, S., Hardy, R.W. and Barrows, F.T. 2001. Digestibility, growth and excretion response of rainbow trout (*Oncorhynchus mykiss* Walnaum) to feeds of different ingredient particle sizes. Aquacult. Res. 32, 885-893

Zimmo, O.R., Al-Saed R. and Gijzen 2000. Comparison between algae-based and duckweed-based wastewater treatment: differences in environmental conditions and nitrogen transformations. Water Sci. Technol. 42(10-11), 215-222.

Chapter Ten
Samenvatting

Samenvatting

Inleiding

In 1990 hadden ongeveer 80 landen, goed voor 40% van de wereldbevolking, te maken met ernstige waterschaarste en het wordt verwacht dat binnen 25 jaar zelfs tweederde van de wereldbevolking zal lijden onder waterschaarste (UNEP, 2002). Extrapolaties volgens het "business as usual" scenario laten zien dat de water consumptie tot 2020 zal toenemen met 40%. Tevens zal er 17% meer water nodig zijn voor voedselproductie om in de behoefte van de groeiende wereldbevolking te voorzien. De landbouw neemt meer dan 70% van alle waterwinning uit rivieren, meren en grondwater voor haar rekening. De geïrrigeerde landbouw zorgt voor 40% van de globale voedselproductie. Ondanks de toename in het percentage mensen dat toegang heeft tot veilig drinkwater van 79% in 1990 tot 82% in 2000, hebben nog steeds 1.1 miljard mensen geen toegang tot veilig drinkwater en 2.4 miljoen mensen hebben geen toegang tot sanitatie. De meeste van hen wonen in Afrika en Azië (UNEP, 2002). De toegang tot sanitatie in rurale gebieden is minder dan in urbane gebieden. Van de mensen zonder sanitatie leeft 80% in rurale gebieden (WHO en UNICEF, 2000). Als gevolg daarvan is er in de laatste decennia een duidelijke noodzaak ontstaan tot het uitbreiden van duurzame en veilige sanitatie naar kleine steden en rurale gemeenschappen. Het doel van nationaal en internationaal beleid was om te identificeren welke opties en randvoorwaarden aanwezig waren voor het introduceren en versneld uitbreiden van dergelijke systemen naar kleine gemeenschappen, met name in waterschaarse gebieden. De Food and Agriculture Organisation (FAO) van de Verenigde Naties (VN) hebben veel aandacht gevestigd op het veilig stellen van de productie van voldoende betaalbaar voedsel voor de bevolking, nu en in de toekomst. De verwachte toename in de vraag naar water vanwege de toename in de wereldbevolking en de per capita water consumptie, zal de geïrrigeerde landbouw negatief beïnvloeden. Ook het verslechteren van bodemvruchtbaarheid vanwege het intensieve gebruik van kunstmest en pesticiden zal de productie van de geïrrigeerde landbouw negatief beïnvloeden. Lozing van communaal afvalwater en drainage water in oppervlaktewater versterkt de eutrofiëring met als gevolg verdere afname van de waterkwaliteit. Verslechtering van de kwaliteit van oppervlaktewater doet de natuurlijke visstand in rivieren en meren afnemen, met als resultaat negatieve gevolgen voor het inkomen van en de voedselveiligheid voor de gemeenschappen in dat gebied. Intensief gebruik van kunstmest en pesticiden doen de vervuiling van het grondwater toenemen. Ernstige vervuiling van oppervlakte en grondwater reduceren de efficiëntie van waterzuiveringssystemen die water innemen vanuit deze vervuilde bronnen. Beter beheer van natuurlijke hulpbronnen door het realiseren van betaalbare afvalwaterzuivering, inclusief nutriënten terugwinning en water hergebruik, kan waterlichamen en bodemvruchtbaarheid beschermen en daarmee de voedselveiligheid voor de samenleving.

Duurzame ontwikkeling is duidelijk gedefinieerd door de FAO (1991) als "het beheer en de bescherming van natuurlijke hulpbronnen en de oriëntatie van technologische en institutionele veranderingen op een manier die het bereiken van menselijke behoeften van huidige en toekomstige generaties verzekert". Zulk een duurzame ontwikkeling is milievriendelijk, technisch en economisch haalbaar en sociaal acceptabel. Conceptueel gezien zijn menselijk afval en drainage water een hulpbron 'in de verkeerde vorm', omdat het in het algemeen in het milieu terecht komt zonder behandeld te zijn. Nutriënten in deze afvalwaters hebben een economische waarde en moeten om zo'n manier gezuiverd worden die duurzaam beheer van natuurlijke hulpbronnen waarborgt. Om dit te verwezenlijken moet afvalwater als een hulpbron gezien worden en de behandeling van afvalwater is een integraal onderdeel van het hele proces van water en nutriënten beheer.

Het concept van gedecentraliseerde sanitatie en hergebruik is een milieuvriendelijke en duurzame optie voor ontwikkelingslanden in aride en semi-aride streken. Anaërobe zuivering van huishoudelijk afvalwater gevolgd door behandeling in eendekroos vijvers is een eenvoudige technologie, met lage kosten, en daarom betaalbaar. Het eendekroos vijver systeem is gebaseerd op zonne-energie en een externe electriciteitsbron is daarom niet nodig. Macrophyten vijvers met eendekroos, en in het bijzonder de rol van eendekroos in het zuiveringsproces en nutriënten terugwinning, is uitgebreid bestudeerd (Al-Nozaily *et al.*, 2000a; Al-Nozaily *et al.*, 2000b, Zimmo *et al.*, 2000, Smith and Moelyowati, 2001; Cheng *et al.*, 2002). In deze vijvers zijn eendekroos en bacteriën de dominante organismen, die het zuiveringsprocess katalyseren. Eendekroos bevat aanzienlijke hoeveelheden eiwit en vertegenwoordigt daarom een waardevol en verkoopbaar bijproduct. Eendekroos is de kleinste bloeiende aquatische plant en heeft geen stengel of blad structuur. Deze unieke eigenschap zorgt ervoor dat eendekroos weinig vezels bevat en het is daarom een goede voedselbron voor Tilapia vis. Daarom ook is eendekroos een veel beter visvoer dan andere aquatische planten, zoals waterhyacint. Waterhyacint bevat relatief weinig eiwit en veel vezels (El-Sayed, 2003). Eendekroos zou ook een beter visvoer kunnen zijn dan oliezaadkoek, omdat dat enige anti-nutritionele factoren bevat (Singh *et al.*, 2003). De toepassing van eendekroos (*Lemna spp.*) in karper polycultuur vijvers had een positief effect op de groei van de vissen (Azim en Wahab, 2003).

Doelstellingen

In dit proefschrift wordt onderzoek beschreven naar een geïntegreerd systeem voor rioolwaterzuivering en hergebruik in aquacultuur, bestaande uit een Upflow Anaerobic Sludge Blanket reactor in combinatie met eendekroos vijvers en visvijvers met tilapia vis. Het UASB-eendekroos-tilapia systeem is een natuurlijk systeem, waarmee een duurzame visteelt mogelijk wordt, omdat rioolwater de enige input van het systeem is. Tot op heden is de geschiktheid van een dergelijk geïntegreerd systeem voor het bevorderen van een duurzame visteelt nog niet bevestigd, omdat tot nog toe weinig onderzoek naar zo'n systeem is uitgevoerd. De algemene doelstelling van dit onderzoek is het evalueren van het systeem van UASB en eendekroosvijvers voor behandeling van huishoudelijk afvalwater en voor productie van visvoer. De specifieke doelstellingen zijn:

-Het bepalen van de efficiëntie van het systeem van UASB en eendekroos vijvers voor behandeling van rioolwater en terugwinning van nutriënten en het evalueren van het effect van temperatuurwisselingen tussen de seizoenen op de zuiveringsefficiëntie en de productie van eendekroos.

-Het bestuderen van het effect van voedergift en van de vorm van het voedsel op de waterkwaliteit en visproductie in vijvers met tilapia, gevoed met eendekroos.

-Een evaluatie van het gebruik van eendekroos als visvoer voor tilapia door het vergelijken van eendekroos met andere soorten visvoer, namelijk tarwezemelen en commercieel visvoer gebaseerd op vismeel.

-Het bestuderen van het effect van waterkwaliteit (leidingwater, gezuiverd rioolwater uit het UASB-eendekroos systeem en voorbezonken rioolwater) op visproductie in tilapia vijvers.

-Het bestuderen van het effect van de microbiologische en andere waterkwaliteitsparameters op de microbiologische besmetting van tilapia geteeld in vijvers met verschillende graden van vervuiling met faecaal materiaal.

-Het bestuderen van het effect van chronische ammoniak toxiciteit op de groei van met eendekroos gevoerde tilapia.

-Het bepalen van verteerbaarheids coefficienten van eendekroos (vers en gedroogd) verwerkt als ingrediënt in visvoer voor intensieve tilapia teelt.

Samenvatting van resultaten en discussie

Hoofdstuk 2 beschrijft het systeem van UASB en eendekroos vijver voor de behandeling van huishoudelijk rioolwater. Het effect van de temperatuur op de efficiëntie van het systeem gedurende de zomer en de winter, alsmede op het lot van stikstof in eendekroos vijvers is bestudeerd. Het experiment werd uitgevoerd in een systeem op pilot schaal, bestaande uit een UASB (40 liter volume) en drie eendekroos vijvers in serie (480 liter elk). De reactor werd gevoed met huishoudelijk rioolwater met een hydraulische verblijftijd (HRT) van 6 uur om dagelijks 160 liter effluent te produceren. Daarvan werd dagelijks 96 liter in de drie eendekroos vijvers gepompt, zodat de HRT 5 dagen in elke vijver was. *Lemna gibba* werd in de vijvers gebracht tot een dichtheid van 600 gram verse biomassa per vierkante meter. Het systeem werd intensief gemonitord gedurende één jaar. De stikstof massabalans werd bepaald voor de drie eendekroos vijvers door het berekenen van de stikstof opname in eendekroos, ammoniak vervluchtiging en stikstof bezinking. Stikstof opname door eendekroos werd berekend door het dagelijks meten van de groeisnelheid van het eendekroos en de samenstelling van het plant materiaal. Ammoniak vervluchtiging werd gemeten door het opvangen van vervluchtigd gas in een 2% boorzuur oplossing waarin vervolgens de ammonium concentratie werd bepaald. Opgehoopt sediment werd geanalyseerd op droge stof, stikstof en fosfor concentratie. Ondanks de significante afname van de UASB efficiëntie voor COD, BOD en TSS verwijdering in de winter (79%, 82%, en 83% voor COD, BOD en TSS respectievelijk in de zomer, en 70%, 73% en 73% in de winter) werd de verwijdering in het totale systeem niet significant beïnvloed (93%, 95% en 91% in de zomer, in vergelijking met 92%, 93% en 91% in de winter). Systeem effluent bevatte 49 mg COD/l, 14 mg BOD/l en 32 mg TSS/l in de zomer, in vergelijking met 73 mg COD/l, 25 mg BOD/l en 31 mg TSS/l in de winter. COD, BOD en TSS verwijderingsefficiëntie in de eendekroos vijvers was hoger in de winter dan in de zomer, vanwege de aanwezigheid van een groot aantal daphnia's en copepoden in het effluent van de vijvers in de zomer. Verwijdering van nutriënten was significant lager gedurende de winter. Tijdens de zomer bevatte het effluent 0.4 mg/l ammonia, 4.4 mg TKN/l en 1.1 mg fosfor/l, met corresponderende verwijderingsefficiënties van 95%, 85% en 78%, respectievelijk. Tijdens de winter waren de waarden voor dezelfde parameters 10.4 mgN/l, 12.2 mgN/l en 2.7 mgP/l, met verwijderings efficiënties van 39%, 53% en 57%, respectievelijk. Dit laat zien dat het proces van nutriëntenverwijdering meer afhankelijk van de temperatuur is dan de verwijdering van BOD en COD. Tijdens de zomer nam het aantal faecale coliformen af van 2.9×10^8 per 100 ml in het rioolwater tot 8.9×10^7 per 100 ml in het UASB effluent en 4.0×10^3 per 100 ml in het systeem effluent. In de winter waren deze waarden 1.1×10^9, 3.2×10^8 en 4.7×10^5 per 100 ml in het rioolwater, UASB effluent en systeem effluent, respectievelijk. De microbiologische kwaliteit van UASB effluent en van rioolwater liet zien dat 63% en 73% van de faecale coliformen werd verwijderd tijdens de zomer en winter, respectievelijk. Dit is in overeenkomst met de 70% verwijdering die gerapporteerd werd door Lettinga *et al.* (1992). De resultaten van de huidige studie laten de effectievere verwijdering van faecale coliformen in eendekroos vijvers tijdens de zomer zien in vergelijking met die in de winter. De verwijderingsmechanismen zouden adsorptie aan eendekroos, sedimentatie, afsterving en grazen door protozoa kunnen zijn.

De stikstof massabalans voor de zomer liet zien dat 80.5% van de totale stikstof verwijdering het gevolg was van opname door het eendekroos, terwijl 4.7% werd verwijderd door sedimentatie en 14.8 % door denitrificatie. Ammoniak vervluchtiging was verwaarloosbaar. Opname van stikstof door eendekroos varieerde tussen 4 en 4.9 kg N/ha/dag in de zomer, en tussen 1.2 en 1.5 kg N/ha/dag in de winter. De totale hoeveelheid fosfor die door het eendekroos teruggewonnen werd variëerde tussen 0.86 tot 0.97 kgP/ha/dag in de zomer, en tussen 0.27 en 0.32 kgP/ha/dag in de winter. Deze resultaten laten het negatieve effect van lage temperaturen op de groei van eendekroos. De eendekroos productie was 126-139 kg droge stof/ha/dag in de zomer en 31-36 kg droge stof/ha/dag in de winter. In het warme seizoen bedroeg de eendekroos productie tot 33 ton droge stof per hectare per 8 maanden (lengte van het warme seizoen). De marktprijs (2002) hiervan zou 16,830 Egyptische ponden kunnen zijn (US$ 4,360), als aangenomen wordt dat de waarde van eendekroos gelijk is aan de lokale prijs voor het goedkoopste visvoer ingrediënt (tarwezemelen). Het droge stof gehalte van eendekroos was 4.3-5.4% met 20-25.7% eiwitgehalte en 0.68-0.9% fosforgehalte. De conclusie van dit hoofdstuk is dat het systeem van UASB en eendekroos technisch geschikt is voor rioolwaterbehandeling in peri-urbane gebieden, kleine stadjes en rurale gebieden, en het levert verkoopbare bijproducten op (eendekroos biomassa).

De voedingswaarde van eendekroos geoogst van eendekroos vijvers voor het voeden van tilapia werd geëvalueerd met behulp van batch experimenten (**Hoofdstuk 3**). Er is onderzocht hoe de hoeveelheid dagelijks toegediend voer en de vorm van toediening van het eendekroos (vers en gedroogd) de groei van Nijltilapia (*Oreochromis niloticus*) en de waterkwaliteit beïnvloedde. Het eendekroos (*Lemna gibba*) werd geëvalueerd als enige voedselbron voor Nijltilapia bij een temperatuur variërend tussen 16 en 25 ^{0}C. Twee batches met jonge vissen met een aanvangsgewicht van ongeveer 8 gram en 20 gram werden getest in visvijvers met 30 liter volume en 0.25 m^2 oppervlakte. De vijvers werden voorzien van vijf vissen per tank en iedere week werd er 15 liter water ververst. Het dagelijkse voerpercentage in de vijf behandelingen was 25% (gedroogd), 50% (gedroogd), 10% (vers), 25% (vers) en 50% (vers), respectievelijk. Het voerpercentage werd berekend op basis van het vers gewicht van zowel het eendekroos als de vis. Het eendekroos dat gebruikt werd in behandelingen 1 en 2 was eerst gedroogd in een oven bij 105 ^{0}C. Het resultaat van de visgroei liet een niet significant (p>0.05) verschil zien tussen de specifieke groeisnelheid (SGR) in de twee batches. De SGRs waren 0.24, 0.24, 0.25, 0.55 en 0.51 in de behandelingen 1, 2, 3, 4, en 5 respectievelijk. Statistische analyse liet een significant (p<0.01) hogere SGR zien voor tilapia gevoed met vers eendekroos (behandelingen 4 en 5) dan voor the parallel behandelingen met gedroogd eendekroos (behandelingen 1 en 2). Dit zou toegeschreven kunnen worden aan het vochtgehalte van vers eendekroos, welke een rol zou kunnen spelen bij het verhogen van de verteerbaarheid van het verse eendekroos. Ook Tacon *et al.* (1995) rapporteerde dat een voerpellet met 39% vochtgehalte een betere resultaat gaf wat beteft SGR en FCR en minder vast afval opleverde dan voerpellets met 12% vochtgehalte, bij het telen van zeebaars (*Lates calcarifer*) en grouper. Guerrero (1980) rapporteerde vergelijkbare resultaten voor kooicultuur van tilapia, waarvoor geobserveerd werd dat het gebruiken van vochtige voerpellets een hogere SGR en netto productie opleverde dan bij het gebruik van droge voerpellets. Bacteriën uit het rioolwater die gehecht zijn aan het eendekroos hebben wellicht ook de microbiële decompositie van eendekroos in de darmen van de vis versneld. Microbiële decompositie van eendekroos in de aanwezigheid van dergelijke bacteriën werd gerapporteerd als drie maal sneller dan zonder deze bacteriën. De tijd die er voor nodig was om de helft van de organische stof in eendekroos af te breken was in de aanwezigheid van bacteriën uit rioolwater een derde van de tijd die ervoor nodig was zonder deze bacteriën (Szabó *et al.*, 2000). Daarbij komt nog dat tijdens het droogproces er sommige essentiële stoffen verloren kunnen gaan en ook dit kan verklaren waarom er een lagere SGR geobserveerd werd in de behandelingen met gedroogd eendekroos. Het relatief grote volume van vers eendekroos had geen negatief effect op de voedselinname, wat verklaard zou kunnen worden door het lage voerpercentage. In het geval van vers eendekroos was de SGR van tilapia met 10% voerpercentage significant lager (p<0.01) dan het geval was voor 25% en 50%. Er was geen significant verschil (p>0.05) tussen 25% en 50% voertoediening. De lage waarde van het optimale voerpercentage (25%) gevonden in dit experiment kan het gevolg zijn van de lage temperatuur (16-25 ^{0}C), die de metabolische activiteit verlaagt omdat de optimale temperatuur voor *Oreochromis niloticus* 28-30 ^{0}C is (Ballerin and Haller, 1982). In het geval van gedroogd eendekroos werd geen significant verschil (p>0.05) gemeten tussen 25% en 50% voerpercentage. De voederconversie (FCR) voor de behandelingen 1 tot 5 waren 4.6, 9.2, 1.7, 1.9 en 4.1 in de batch met 20 grams tilapia en 4.4, 8.7, 1.6, 1.9 en 4.2 in de batch met 8 grams tilapia, respectievelijk. De eiwit omzettingsefficiëntie (PER) was 0.9, 0.45, 2.5, 2.1 en 1.0 in de behandelingen 1, 2, 3, 4 en 5, respectievelijk. De resultaten van de stikstof massabalans lieten zien dat het stikstof gehalte van gedroogd sediment (vast afval) variëerde tussen 4% en 10.5%. In het geval van het optimale voerpercentage (25% vers eendekroos) was de waarde 4.5%. Deze hoeveelheid stikstof is gelijk aan 8% van de stikstof in het voer, terwijl stikstof opname door de vissen 26-28% uitmaakte van de totale stikstof input. Het grootste gedeelte van het stikstof verlies was het gevolg van effluent lozing (10% en 25% voerpercentage) of het gevolg van ophoping in het sediment (50% voerpercentage).

Het gebruik van zowel gezuiverd rioolwater als van eendekroos biomassa van het systeem van UASB en eendekroos vijvers werd geëvalueerd voor de teelt van Nijltilapia (*Oreochromis niloticus*) in visvijvers (**Hoofdstuk 4**). Een vergelijking werd uitgevoerd tussen eendekroos geteeld op anaëroob gezuiverd rioolwater en een lokaal gebruikt visvoer ingrediënt, tarwezemelen, als twee verschillende aanvullende voeders. Drie verschillende soorten water (leidingwater, effluent van het UASB-eendekroos systeem en voorbezonken rioolwater) werden onderzocht. Het experiment werd uitgevoerd met behulp van 5 tanks, ieder met 1 m^2 oppervlak. Elke tank werd voorzien van 10 jonge Nijltilapia van ongeveer 20 gram. De voedingswaarde van eendekroos werd vergeleken met tarwezemelen door beiden apart toe te passen als enige voedingsbron, terwijl gezuiverd rioolwater of leidingwater als waterbron werd gebruikt. Het eendekroos voerpercentage werd vastgesteld op 25 gram vers eendekroos per 100 gram vers vis-gewicht. De tanks gevoed met tarwezemelen ontvingen dezelfde hoeveelheid droge stof als de eendekroos tanks. Parralel aan deze incubaties ontving een andere tank voorbezonken rioolwater als de enige voedingsbron. De resultaten voor groeisnelheden lieten zien dat voor tanks met leidingwater de SGR voor tilapia gevoerd met vers eendekroos significant hoger (p<0.01) was dan de SGR voor de tilapia gevoerd met tarwezemelen. Dit zou kunnen komen door het significant hogere eiwitgehalte in eendekroos in vergelijking tot tarwezemelen. Er was geen significant verschil (p>0.05) tussen de behandelingen in het geval van de tanks gevoed met gezuiverd rioolwater. Dit zou kunnen komen doordat afpaaien in de tanks met eendekroos eerder optrad dan in de tanks met tarwezemelen (75 dagen later). Dit zou de SGR negatief beïnvloed kunnen hebben met als resultaat geen significant verschil met de tanks gevoed met tarwezemelen. In de tanks gevoed met tarwezemelen was de SGR van tilapia geteeld in gezuiverd rioolwater (TWP) significant hoger (p<0.01) dan de SGR van de tilapia geteeld in de tanks met leidingwater (FWP). In de tanks gevoed met eendekroos was de SGR van tilapia geteeld in gezuiverd rioolwater (TDP) significant lager (p<0.05) dan de SGR van de tiliapia in de tanks met leidingwater (FDP). Dit kan toegeschreven worden aan het vroege afpaaien in TDP, zoals al eerder beschreven. Netto visproducties waren 11.8, 9.6, 8.9 en 6.4 ton/ha/jaar in TDP, FDP, TWP en FWP. De laagste netto visproductie was in de tank gevoed met voorbezonken rioolwatr (SSP) en was –0.16 ton/ha/jaar. Het slechte resultaat in de SSP wordt toegeschreven aan de hoge mortaliteit in deze tank, die 60% in de volwassen vissen en 38% in de jongbroed bedroeg. Deze mortaliteit zou het gevolg kunnen zijn van chronische NH$_3$ toxiciteit. De NH$_3$ concentratie was gemiddeld 0.12 mgN/l tijdens het experiment. Vissterfte werd geobserveerd tijdens de laatste twee maanden van het experiment (herfst), toen de gemiddelde NH$_3$ concentratie 0.45 mgN/l was. De visproductie in tanks gevoed met gezuiverd rioolwater was hoger dan in de tanks gevoed met leidingwater. Dit kan toegeschreven worden aan het effect op de primaire productie van de nutriënten die aanwezig zijn in gezuiverd rioolwater, namelijk het voorzien in een deel van de nutriënten behoefte van de vis (Tidwell *et al.*, 1995; Agustin, 1999). De aanwezigheid van dapnia en copepoden in het gezuiverde rioolwater kan ook bijgedragen hebben aan de hoge productie in de tanks met gezuiverd rioolwater (McEvoy *et al.*, 1998; Warburton *et al.*, 1998; Kibria *et al.*, 1999; Lienesch en Gophen, 2001).

In **hoofdstuk 5** wordt onderzoek beschreven naar de effecten van langdurige blootstelling aan niet accuut dodelijke NH$_3$ concentraties op de groei van jonge tilapia (*Oreochromis niloticus*) gevoed met vers eendekroos (*Lemna gibba*). Het experiment werd uitgevoerd in duplo gedurende 75 dagen met jonge tilapia met een gemiddeld gewicht van 20 gram. Vijf nominale totale ammonium concentraties (controle, 2.5, 5, 7.5 en 10 mg N/liter) vormden de behandelingen. Statistische analyse van de specifieke groeisnelheden (SGR) liet een niet significant (p>0.05) verschil zien tussen de SGR (0.71) van tilapia in de controle (0.004 mg NH_3-N/liter) en de SGR (0.67) van tilapia blootgesteld aan 0.068 mg NH_3-N/liter. De SGR van tilapia blootgesteld aan meer dan 0.068 mg NH_3-N/liter (0.144, 0.262 en 0.434 mg NH_3-N/liter) was significant minder (p<0.01). De maximale concentratie zonder effect was 0.068 mg NH_3-N/liter, terwijl de laagste concentratie waarbij een effect werd geobserveerd 0.144 mg NH_3-N/liter was. Een toename in de NH_3 concentratie had tot gevolg een toename in de voederconversie (FCR). Bij een concentratie van 0.144 mg NH_3-N/liter was de FCR 1.6 maal de de waarde geobserveerd in de controle, terwijl bij 0.262 mg NH_3-N/liter de waarde 2.7 maal hoger was. De eiwit omzettingsefficiëntie (PER) werd ook negatief beïnvloed door NH_3 concentraties hoger dan 0.068 mg NH_3-N/liter. Bij 0.434 mg NH_3-N/liter was de FCR toegenomen met een factor 4.3. De huidige studie concludeerde dat voor het telen van *Oreochromis niloticus* in visvijvers, de NH_3 lager moet zijn dan 0.1 mg NH_3-N/liter. In dit experiment werd geen vissterfte geconstateerd, zelfs niet bij de hoogste concentratie (0.43 mg NH_3-N/liter). Dit resultaat schijnt in tegenspraak te zijn met de data beschreven in hoofdstuk 4, waar vissterfte gerapporteerd werd in de tank gevoed met voorbezonken rioolwater bij een NH_3 concentratie van 0.45 mg NH_3-N/liter. Dit kan toegeschreven worden aan de cumulatieve negatieve effecten van andere waterkwaliteitsparameters, zoals de aanwezigheid van bacteriën uit rioolwater en de lage zuurstofconcentratie in de vroege morgen in de tanks gevoed met voorbezonken rioolwater. Deze parameters zouden een synergistisch effect kunen hebben op de toxiciteit van NH_3 in dergelijke tanks.

In **hoodstuk 6** wordt onderzoek beschreven naar de microbiologische kwaliteit van tilapia geteeld in vier met faecaal materiaal verontreinigde visvijvers, zoals ook beschreven in hoofdstuk 4. De TDP vijvers werden verontreinigd vanuit twee bronnen, namelijk gezuiverd rioolwater en vers eendekroos geteeld op anaëroob gezuiverd rioolwater. In de TWP en FDP vijvers was er maar één bron van besmetting, namelijk gezuiverd rioolwater in het eerste geval en vers eendekroos in het tweede geval. De laatste vijver (SSP) werd alleen gevoed met voorbezonken rioolwater als water bron en indirecte voeding. Het gezuiverde rioolwater bevatte gemiddeld 4 x 10^3 faecale coliformen per 100 ml, terwijl het gemiddelde aantal in voorbezonken rioolwater 2.1 x 10^8 faecale coliformen per 100 ml was. Het aantal faecale coliformen gehecht aan het eendekroos variëerde van 4.1 x 10^2 tot 1.6 x 10^4 faecale coliformen per gram vers gewicht. Monitoren van de microbiologische waterkwaliteit toonde een gemiddeld aantal faecale coliformen aan van 2.2 x 10^3, 1.7 x 10^3, 1.7 x 10^2 en 9.4 x 10^3 per 100 ml in TDP, TWP, FDP en SSP, respectievelijk. Statistische analyse liet een niet significant (p>0.05) verschil in faecale coliform concentraties zien tussen de twee vijvers met gezuiverd rioolwater (TDP en TWP) en de SSP vijver. De concentratie faecale coliformen in de FDP vijver was significant lager dan in de vijvers gevoed met gezuiverd rioolwater (p<0.01) en de SSP vijver (p<0.05). Bepaling van de microbiologische kwaliteit van tilapia geteeld in deze vijvers liet zien dat alle monsters van weefsels en organen besmet waren met faecale coliformen, behalve het spierweefsel, dat niet faecaal besmet was. Het rangschikken van de organen en weefsel naar besmettingsgraad liet een afname in besmettingsgraad zien in de volgorde darmkanaal > kieuwen > huid > lever. In het geval van de TDP vijver was er geen significant verschil (p>0.05) in microbiologische kwaliteit tussen het darmkanaal en de kieuwen, noch tussen de huid en de lever. In het geval van de TWP vijver werd een significant hogere (p<0.05) faecale besmetting gemeten in het darmkanaal in vergelijking met de kieuwen, maar er was geen verschil tusen de huid en de lever (p>0.05). In de FDP vijver werd een significant hogere (p<0.01) besemetting gemeten in het darmkanaal, in vergelijking tot de andere organen. In de SSP vijver was er geen significant verschil tussen het darmkanaal en de andere organen (p>0.05), maar de besmetting van deze organen was hoger dan die van de lever en de kieuwen (p<0.05). Ondanks de afwezigheid van een significant verschil tussen de microbiologische kwaliteit van het water in de SSP vijver en de vijvers met gezuiverd rioolwater, werd een significant hogere besmetting van vis organen in deze laatste vijver gemeten. Dit kan toegeschreven worden aan andere waterkwaliteits parameters, zoals NH_3 en nitriet, die significant hoger waren in de SSP vijver. Deze slechte waterkwaliteit in de SSP vijver had tot gevolg dat de besmetting met faecale coliformen van vis organen ongeveer één log_{10} eenheid hoger was dan in de vijvers met betere waterkwaliteit. Zuivering van rioolwater wordt daarom aangeraden voordat het gebruikt wordt in de visteelt, niet alleen voor verwijdering van faecale coliformen, maar ook voor verwijdering van NH_3 en nitriet.

In **hoofdstuk 7** wordt vers eendekroos vergeleken met een lokaal, op vismeel gebaseerd visvoer, met het gebruik van gezuiverd rioolwater en gede-chloreerd leidingwater als waterbron. Vers eendekroos (*Lemna gibba*) werd dagelijks geoogst van de stabilisatie vijvers met eendekroos en gebruikt in een cultuur met Nijltilapia (*Oreochromis niloticus*) met een gemiddeld aanvangsgewicht van 23 gram. Het dagelijkse voerpercentage in de vijvers gevoerd met eendekroos was 25 gram vers eendekroos per 100 gram vers vis gewicht. De visvijver gevoerd met het commerciële voer ontving dezelfde hoeveelheid eiwit als de vijvers gevoerd met eendekroos. Het experiment werd uitgevoerd in vier vijvers, elk met 1 m^2 oppervlak en 480 liter volume, bij een temperatuur van 26-34 ^0C gedurende een groeiperiode van 120 dagen. In de vijvers met leidingwater groeiden de jonge tilapia vissen die het commerciële voer kregen significant sneller (p<0.01) dan de tilapia die eendekroos gevoerd kregen. Dit kan toegeschreven worden aan deficiëntie in een aantal essentiële voedingsstoffen in eendekroos, zoals cysteïne en methionine (Hammouda *et al.*, 1995). Deze esentiële voedingsstoffen zouden in het geval van het gebruik van gezuiverd rioolwater aangevuld kunnen worden via de natuurlijke productiviteit. Inderdaad werd in deze vijvers geen significant verschil tussen commercieel voer en eendekroos waargenomen. D'Abramo en Conklin (1995) rapporteerde dat de natuurlijke biota significant bij kan dragen aan de behoefte aan somige voedingsstoffen, zoals vitaminen, cholesterol, phospholipiden en mineralen. Dit werd ook bevestiged door de resultaten van Omondi *et al.* (2001). De daphnia en copepoden die waargenomen werden in gezuiverd rioolwater kunnen ook deficiënties van het eendekroos compenseren, omdat deze crustaceans een zeer goede voedingswaarde voor visteelt bezitten (Bogatova *et al.*, 1973; Proulx en de La Noue, 1985). Deze positieve effecten van gezuiverd rioolwater zijn ook waargenomen in vergelijking met het gebruik van leidingwater. De SGR en visproductie waren hoger in de vijvers gevoed met gezuiverd rioolwater, zowel bij de vijvers met eendekroos als commercieel voer als input. Netto visproductie was 15.7, 16.8, 8.5 en 11.2 ton/ha/jaar in TDP, TCP, FDP en FCP, respectievelijk.

In **hoofdstuk 8** worden de metingen van de verteerbaarheidscoëfficiënten voor eendekroos (*Lemna* gibba) als voer voor Nijltilapia (*Oreochromis niloticus*) beschreven. Tilapia werd gevoerd met 4 verschillende voeders (de behandelingen) met gelijk stikstofgehalte. Daarnaast werd een controle behandeling uitgevoerd met een contole voer. Het controle voer bevatte vismeel (35%), maismeel (29%), korenmeel (20%), tarwezemelen (10%), visolie (3%), diamol (2%) en premix (1%). Het effect van gedeeltelijke vervanging van het controle diëet met 20% en 40% gedroogd eendekroos (percentages gebaseerd op droge stof) in de voeders voor behandeling 1 (T1) en behandeling 2 (T2) werd onderzocht. In behandeling 3 (T3) en 4 (T4) werd vers eendekroos gebruikt voor het vervangen van respectievelijk 20% en 40% van het controle diëet (percentages gebaseerd op droge stof). Gedurende de 49 dagen van het experiment werd het faecale materiaal van de vissen verzameld. Tevens werd de waterkwaliteit gemonitord. Aan het eind van de experimentele periode werden groeisnelheid, verteerbaarheids coëfficiënten, lichaamssamenstelling, percentage opgevangen faecaliën en de stikstof massabalans bepaald. De specifieke groeisnelheid van tilapia was respectievelijk 1.51 ± 0.07, 1.38 ± 0.03, 1.31 ± 0.06, 1.44 ± 0.02 en 1.33 ± 0.05 in de controle behandeling en behandelingen 1 tot 4. Er was geen significant verschil tussen de SGR van de controle en behandeling T3, terwijl de andere behandelingen een significant lagere groeisnelheid lieten zien. Alle behandelingsdiëten lieten goede waarden voor de FCR (0.98-1.1) en voor de PER (2.49-2.78) zien. De gemiddelde waarde van FCR en PER voor het controle dieet was respectievelijk 0.91 en 3. Alleen voor T2 werd een significant ($p<0.05$) verschil met de controle voor het percentage opgevangen faecaliën gemeten. De hoeveelheid opgevangen faecaliën in T1 was significant ($p<0.05$) lager dan in de andere behandelingen. Er was geen significant verschil tussen het droge stof gehalte van de faecaliën, behalve in behandeling T4 waarin een significant lagere waarde werd gevonden dan in de controle en behandeling T2. De verteerbaarheid van de droge stof (DMD) variëerde van 61.8% in T4 tot 85.2% in de controle en die voor de behandelingen variëerden tussen 64% en 71%. Alle behandelingsdiëten hadden een hoge eiwit verteerbaarheid (78%-92%) en een hoge energie verteerbaarheid (78.1%-90.7%). Het is daarom mogelijk om eendekroos te gebruiken als vervanger van vismeel en andere plantaardige ingrediënten in een op vismeel gebaseerd diëet, bij een vervanging van 20% en 40%. De verteerbaarheidscoëfficiënten van gedroogd en vers eendekroos hadden een DMD van respecievelijk 66.8, 63.3, 45.8 en 28.3 in de behandelingen 1 tot 4. Deze resultaten, zelfs in het geval van 40% vers eendekroos, zijn vergelijkbaar met de gerapporteerde waarden (27.5%-50%) voor tarwegrint, korenmeel, maismeel en gele mais (Sullivan en Reigh, 1995; Fagbenro, 1996). Eiwit verteerbaarheidscoëfficiënten (PDC) voor eendekroos waren 78.4, 79.9, 77.6 en 75.9, terwijl de energie verteerbaarheid waarden (ED) 59.8, 60.9, 64.5 en 58.4 waren in T1, T2, T3 en T4, respectievelijk. Voor de behandelingen met 20% vervanging door eendekroos was er geen significant verschil tussen gedroogd en vers eendekroos wat betreft de verteerbaarheidscoëfficiënten voor droge stof, eiwit, vet, fosfor en energie. Echter, voor de behandeling met 40% vervanging door eendekroos had vers eendekroos een signficant lagere verteerbaarheidscoëfficiënten voor droge stof, eiwit, vet en fosfor. Deze resultaten kunnen in het geval van 40% vervanging toegeschreven worden aan verdunning van de enzyme-concentratie door vocht afkomstig uit het verse eendekroos. Het malen van het gedroogde eendekroos kan ook een positief effect gehad hebben op de verteerbaarheid van droog eendekroos, omdat het kapot maken van de celwand van eendekroos zorgt voor het blootstellen van de inhoud aan enzymen. Het verminderen van de deeltjesgrootte van eendekroos zorgt ook voor een toename van de specifieke oppervlakte beschikbaar voor enzymreacties, met als resultaat een toename van de verteerbaarheidscoëfficiënten. Lichaamssamenstelling van tilapia liet zien dat tilapia gevoed met eendekroos een significant ($p<0.05$) hogere verhouding van fosfor en eiwit heeft en een lager ($p<0.05$) vetgehalte. Dit kan aantrekkelijk zijn voor consumenten.

In hoofdstuk 3 is beschreven dat de groeisnelheid van tilapia gevoed met vers eendekroos hoger was dan van tilapia gevoed met gedroogd eendekroos. De resultaten voor de verteerbaarheidscoefficient voor gedroogd en vers eendekroos in dit hoofdstuk schijnen in tegenspraak te zijn met die in hoofdstuk 3. Dit kan verklaard worden door de volgende redenen. Ten eerste kunnen de positieve effecten van het maalproces (Zhu *et al.*, 2001) in de bepaling van de verteerbaarheid de negatieve effecten van het drogen (zoals beschreven in hoofdstuk 3) gecompenseerd hebben. Ten tweede, in hoofdstuk 3 was de gebruikte eendekroos het relatief grote *Lemna gibba* en in hoofdstuk 8 het kleine *Lemna minor*. De engiszins grotere afmetingen van *Lemna gibba* maken het moeilijker voor tilapia om het plant materiaal door te slikken zonder het te kouwen, terwijl het kleine *Lemna minor* uit hoofdstuk 8 makkelijk doorgeslikt kan worden zonder kouwen. Het kouwen doet de verteerbaarheid toenemen door het kapot maken van de celwand van de plant. Ten derde kunnen de bacteriën die gehecht waren aan *Lemna gibba* (hoofdstuk 3) een positief effect gehad hebben op de verteerbaarheid.

Conclusies

De efficiëntie van een systeem van UASB en eendekroos vijvers voor de verwijdering van COD, BOD en TSS was voldoende gedurende het hele jaar, terwijl nutriënten terugwinning in eendekroos vijvers significant lager was bij lagere temperaturen (13-20 ^{0}C) omdat nutriënten terugwinning een proces is wat gefaciliteerd wordt door groei van eendekroos. De verwijdering van faecale coliformen in eendekroos vijvers werd negatief beïnvloed door de afname in temperatuur, doordat de coliform verwijderingsmechanismen beïnvloedt worden door temperatuur, beschikbaarheid van nutriënten en de mate waarin eendekroos geoogst wordt. In het algemeen kan gezegd worden dat het UASB-eendekroos systeem technisch geschikt, sociaal acceptabel, duurzaam en milieuvriendelijk is voor huishoudelijk afvalwaterzuivering in rurale gebieden en kleine gemeenschappen.

Het gebruik van het UASB-eendekroos vijver systeem voor een geïntegreerd systeem van rioolwaterzuivering en aquacultuur is beter en veiliger dan het directe gebruik van rioolwater voor visteelt. Vers eendekroos (*Lemna gibba*) is beter (grotere visproductie) dan tarwezemelen als aanvullend voer en heeft dezelfde voedingswaarde als diëten gebaseerd op vismeel, vooral in vijvers gevoed met gezuiverd rioolwater. Het gebruik van gezuiverd huishoudelijk rioolwater van het UASB-eendekroos systeem, als vervanging van leidingwater, is duurzaam omdat bemesting van tilapia vijvers met gezuiverd rioolwater uit het syteem de visproducite significant verhoogt zonder een toename in de inzet van natuurlijke hulpbronnen.

Waterkwaliteitsparameters zoals NH_3 en nitriet in visvijvers kan de mate van bacteriële besmetting van visorganen beïnvloeden, waarschijnlijk door the effect van de waterkwaliteit op het immuunsysteem van tilapia. Het wordt daarom aangeraden om rioolwater te zuiveren voor het te gebruiken in de visteelt. De lage zuurstof concentratie en microbiologische waterkwaliteit zou een synergistisch effect kunnen hebben op de chronische NH_3 toxiciteit. Om negatieve effecten van NH_3 op de groei van tilapia te voorkomen moet de NH_3 concentratie in vijvers lager blijven dan 0.1 mg NH_3-N/liter.

Verteerbaarheidscoëfficiënten van eendekroos (*Lemna minor*) zijn beter dan voor een aantal conventionele plantaardige ingrediënten die gebruikt worden in visvoer. De resultaten van de verteerbaarheidsstudie met diëten met 20% en 40% gedroogd eendekroos zijn een stimulans voor het gebruik van gedroogd eendekroos in de productie van visvoer. Het gebruik van eendekroos in verse of gedroogde vorm kan kostbare en vaak slecht beschikbare diëten gebaseerd op vismeel vervangen.

Aanbevelingen

De nutriëntenverwijdering en terugwinning door het UASB-eendekroos systeem is significant lager bij lagere temperaturen (13-20 ^0C). Daarom wordt het opslaan van effluent in een reservoir gedurende de winter aanbevolen om in de zomer te worden hergebruikt in irrigatie of aquacultuur. In het geval van hergebruik gedurende de winter in de landbouw of aquacultuur moet speciale aandacht gegeven worden aan de concentraties ammonium en Kjeldahl stikstof.

Referenties

Agustin, C.P. 1999. Role of natural productivity and artificial feed in the growth of freshwater prawn, *Macrobrachium borelli* (Nobili, 1896) cultured in enclosures. J. Aquacult. Trop. 14(1), 47-56

Al-Nozaily, F., Alaerts, G. and Veenstra, S. 2000. Performance of duckweed-covered sewage lagoon. 1. Oxygen balance and COD removal. Water Res. 34(10), 2727-2733

Al-Nozaily, F., Alaerts, G. and Veenstra, S. 2000a. Performance of duckweed-covered sewage lagoon. 2. Nitrogen and phosphorus balance and plant productivity. Water Res. 34(10), 2734-2741

Azim, M.E. and Wahab, M.A. 2003. Development of a duckweed-fed carp polyculture system in Bangladesh. Aquaculture 218, 425-438

Balarin, J.D. and Haller, R.D. 1982. The intensive culture of tilapia in tanks, raceways and cages. In: James F. Muir & Ronald J. Roberts (Eds.), Recent advances in aquaculture. Croom Helm Ltd, London. pp. 265-355.

Bogatova, I.B., Schhesbina, M.A., Ovinnikova, V.V. and Tagisova, N.A. 1973. The chemical composition of certain planktonic animals under different growing conditions. Hydrobiol. J. 7, 39-43

Cheng, J., Bergamann, B. A., Classen, J.J., Stomp, A.M. and Howard, J.W. 2002. Nutrient recovery from swine lagoon water by *Spirodela punctata*. Biores. Technol. 81, 81-85.

D'Abramo, L.R. and Conklin, D.E. 1995. New development in the understanding of the nutrition of penaeid and caridean species of shrimp. Proceedings special session on shrimp farming, aquaculture 95, World Aquaculture Society, Louisiana, USA, pp. 95-107

El-Sayed, A.M. 2003. Effects of fermentation methods on the nutritive value of water hyacinth for Nile tilapia *Oreochromis niloticus* (L.) fingerlings. Aquaculture 218, 471-478

Fagbenro, O.A. 1996. Apparent digestibility of crude protein and gross energy in some plant and animal-based feedstuffs by *Clarias isheriensis* (Siluriformes: Clariidae) (Sydenham 1980). J. Appl. Ichthyol. 12, 67-68

Guerrero, R.D.III, 1980. Studies on the feeding of *Tilapia nilotica* in floating cages. Aquaculture, 20, 169-175

Hammouda, O., Gaber, A. and Abdel-Hameed, M.S. 1995. Assessment of the effectiveness of treatment of wastewater-contaminated aquatic systems with *Lemna gibba*. Enz. and Micro. Technol. 17: 317-323

Kibria, G., Nugegoda, D., Fairclough, R., Lam, P. and Bradley, A. 1999. Utilization of wastewater-grown zooplankton: Nutritional quality of zooplankton and performance of silver perch bidyanus bidyanus (Mitchell 1838) (Teraponidae) fed on wastewater-grown zooplankton. Aquacult. Nutr. 5, 221-227

Lettinga, D., de Man, A., van der Last, A. R. M., Wiegan, W., van Knippenberg, K., Frings, J. and van Bauren, J. C. L., 1992. Anaerobic treatment of domestic sewage and wastewater. "Proceeding of first Middle East conference on water supply and sanitation for rural areas". Pp. 164-170, Cairo, Egypt.

Lienessch, P.W. and Gophen, M. 2001. Predation by inland silversides on exotic cladoceran, daphnia lumholtzi, in lake Texoma, USA. J. Fish Biol. 59, 1249-1257

McEvoy, L.A., Naess, T., Bell, J.G. and Lie, Ø. 1998. Lipid and fatty acid composition of normal malpigmented Atlantic halibut (*Hippoglossus hippoglossus*) fed enriched Artemia: A comparison with fry fed wild copepods. Aquaculture 163, 237-250

Omondi, J.G., Gichuri, W.M. and Veverica, K. 2001. A partial economic analysis for Nile tilapia (*Oreochromis niloticus* L.) and sharptoothed catfish *Clarias gariepinus* (Burchell 1822) polyculture in central Kenya. Aquacult. Res. 32, 693-700

Proulx, D. and De La Noue, J. 1985. Harvesting *Daphnia magna* grown on urban tertiarily treated effluents. Water Res. 19(10), 1319-1324.

Singh, K., Garg, S.K., Kalla, A. and Bhatnagar, A. 2003. Oilcakes as protein sources in supplementary diets for the growth of *Cirrhinus mrigala* (Ham.) fingerlings: laboratory and field studies. Biores. Technol. 86, 283-291

Smith, M.D. and Moelyowati, I. 2001. Duckweed based wastewater treatment (DWWT): design guidelines for hot climates. Water Sci. Technol. 43(11), 291-299

Sullivan, J.A. and Reigh, R.C. 1995. Apparent digestibility of selected feedstuffs in diets for hybrid stripped bass (*Morone saxatilis* × *Morone chrysops*). Aquaculture, 138, 313-322

Szabó, S., Braun, M., Nagy, P., Balazsy, S. and Reisinger, O. 2000. Decomposition of duckweed (*Lemna gibba*) under axenic and microbial conditions: flux of nutrients between litter water and sediment, the impact of leaching and microbial degradation. Hydrobiologia 434, 201-210.

Tacon, A.G.J., Phillips, M.J. and Barg, U.C. 1995. Aquaculture feeds and the environment: the Asian experience. Water Sci. Technol. 31(10), 41-59.

Tidwell, J.H., Webster, C.D., Sedlacek, J.D., Weston, P.A., Knight, W.L., Hill, Jr S.J., D'Abramo, L.R., Daniels, W.H., Fuller, M.J. and Montanez, J.L. 1995. Effects of complete and supplemental diets and organic pond fertilisation on production of *Macrobracheium rosenbergii* and associated benthic macroinvertebrate populations. Aquaculture, 138, 169-180

UNEP 2002. State of the environment and policy retrospective: 1972-2002. In Global Environment Outlook 3 Past, Present and Future Perspectives (Eds. Production team of UNEP), UNEP, London, UK.

Warburton, K., Retif, S. and Hume, D. 1998. Generalists as sequential specialists: diets prey switching in juvenile silver perch. Environ. Biol. Fishes 51, 445-454

WHO and UNICEF 2000. Global Water Supply and Sanitation Assessment 2000 Report, pp 80

Zhu, S., Chen, S., Hardy, R.W. and Barrows, F.T. 2001. Digestibility, growth and excretion response of rainbow trout (*Oncorhynchus mykiss* Walnaum) to feeds of different ingredient particle sizes. Aquacult. Res. 32, 885-893

Zimmo, O.R., Al-Saed R. and Gijzen 2000. Comparison between algae-based and duckweed-based wastewater treatment: differences in environmental conditions and nitrogen transformations. Water Sci. Technol. 42(10-11), 215-222.

Acknowledgements

First of all I would like to thank my god for giving me a good mind to follow the right way that lighten for me. My warmest thanks to my country, Egypt for giving me and all of the Egyptian people the opportunity to learn and know our history. Great thanks to all members of my family, parents, sisters and brothers; special thanks to my mother without her I was not be able to complete my education.

I would like to express my great gratitude to, Prof. Dr. Huub Gijzen, my promoter for his support, encouragement and fruitful discussions and comments and for reviewing the final manuscript. My gratitude goes to my promoter Prof. Dr. Fatma El-Gohary for giving me this opportunity to work within the framework of the "Wasteval" project. Great thanks to Prof. Dr. Fayza Nasr for her support and continuous encouragement; she really gave me a big push when I needed it. I am greatly and deeply indebted to my co-promoters Prof. Dr. Johan Verreth and Dr. Peter van der Steen for their support, assistance and reviewing of the manuscript. I am also grateful to Dr. Jules van Lier, the Wasteval project manager who made a lot of efforts to provide me with a comfortable stay in the Netherlands; his help and friendship in difficult times will never be forgotten. Great thanks to my Egyptian colleagues at IHE for their continuous encouragement, Mousa, Salih and all of them.

It is my pleasure to remember the staff members of the fish culture and fisheries group of the Wageningen University, Dr. Johan Schrama and the entire group there. Last but not least, I would like to express my great thanks to the staff members of the Water Pollution Control Department of the National Research Centre (NRC), Egypt. Special thanks to the technicians of the wastewater treatment group, Kamal Hassan, Fatma Refaat, Hoda Hegazy, Ahmed Ali and my colleague Mohamed Ali.

Curriculum Vitae

Saber Abdel-Aziz Abdel-Salam Mohammed El-Shafai, the author of this thesis, was born in February 24th, 1970, in the city of Oseim, Giza, Egypt. His elementary studies were completed at the Oseim preparatory school in 1985. In May 1988 he graduated from the Oseim higher degree secondary school, at the scientific department. After that, he joined the faculty of science, Cairo University. He was awarded his B. Sc. with a general degree "very good", from the department of Zoology in May 1992.

After finishing his military service in January 1994, he attended post-graduate courses as partial fulfillment of the requirements for the M. Sc. Degree. He completed successfully these courses in October 1995 from the sub-department of Histology, Cytology and Genetics, Zoology Department, Faculty of Science, Cairo University. In January 1995 he was awarded a scholarship from the Egyptian Academy of Scientific Research for two years to complete the practical work of his M. Sc. in the department of Water Pollution Research and Control, the National Research Center, Egypt. During the stay at this Center, he was chosen to work permanently in the Department of Water Pollution Research and Control as research assistant.

In June 1999, he was awarded the M. Sc. Degree "Cytology, Histology and Genetic" at the Zoology Department, Faculty of Science, Cairo University. His M. Sc. Title was " Wastewater treatment and Reuse in Fish Aquaculture". After getting his M. Sc. Degree, he ranked a position of assistant researcher in his department. He has been participating in several research projects, both national and international, that dealt with the treatment of domestic and industrial wastewater using different technologies, conventional and non-conventional. Within the framework of the "Wasteval" project he received a sandwich scholarship to complete his Ph.D. program at the UNESCO-IHE institute for water education, in cooperation with Wageningen University and Research, Department of Environmental Technology, The Netherlands.

Address:
Saber Abdel-Aziz Abdel-Salam Mohamed El-Shafai
Water Pollution Control Department
The National Research Center
El-Tahrir Street, Dokki, Giza, Egypt.
Email: Saberabdelaziz@hotmail.com

Financial support

The Dutch government financially supported this research study within the framework of Sail-IOP/SPP (WASTEVAL) Project. Sail-IOP/SPP "WASTEVAL" Project is collaboration project between Water Pollution Control Department, NRC, Egypt and UNESCO-IHE Institute for Water Education and Wageningen University and Research, The Netherlands.

Printed and bound by CPI Group (UK) Ltd, Croydon, CR0 4YY

23/10/2024

01777685-0008